The wonder of creation and the enigma of spiritual beings

Words of Light and Life
VOLUME 5

DR F P MÖLLER

J L van Schaik
RELIGIOUS BOOKS

Published by J L van Schaik Publishers
1064 Arcadia Street, Hatfield, Pretoria

First edition 1997
ISBN 0 627 02198 0

Cover design by Eddie Fullhard
Typesetting in 10½ on 12 pt News Gothic by
Sonja Reinecke, Studio S
Printed and bound by National Book Printers
Drukkery Street, Goodwood, Western Cape

Table of contents

CHAPTER 1 Protology: Biblical doctrine of creation

Protology deals with the doctrine of first things as contained in the biblical account of creation. It deals with matters such as: the origination of heaven and earth; its subsequent development; the origination of living beings and in particular, the creation of man; the impact of the Fall on creation; the pronouncements of science concerning the origin of all things and God's purpose with creation. It may be more logical, having commenced dogmatic study with protology, to end eventually with eschatology (the doctrine of the last things). However, the beginning of all things can only be understood after an overview of certain key truths revealed in the Bible.

1 CREATION AS COMMUNICATED BY GENESIS

In Genesis 1 and 2 we read how creation took place and how everything came into existence in the beginning. However, before the biblical creation narrative is studied, we firstly refer to other ancient stories about creation.

1.1 The creation narrative of the Bible and other creation narratives

There is good reason to believe that mankind did not learn about the creation events only after the Book of Genesis came into existence. The fact that Adam and Eve communicated with the Lord God in the garden of Eden, implies that they generally knew how they and everything around them, had come into being. This is indicated by the reason given in the Decalogue for the keeping of the seventh day as the Sabbath. The keeping of the Sabbath is based on God's creation of all things in six days, after which he rested on the seventh day. Since the legislation in connection with the Sabbath took place before the writing of the Pentateuch, it should be accepted that at the time of the giving of this law the Israelites already knew something about the creation narrative.[1] At the time when the biblical account of creation was written, there were in fact already other creation narratives in circulation.

1 G Ch Aalders states the following: "The way the fourth commandment was given is proof that God, right from the beginning, had revealed to mankind knowledge about his work of creation. In that commandment it is mentioned, as an acknowledged fact, that the creation of heaven and earth, the sea and all that are in it, took place in six days, after which God rested on the seventh day. We have to note: these things were not made known on the occasion of the pronouncement of the law with its ten commandments; it was rather assumed as a fact, and therefore was presented as the ground and urgent reason for Israel to rest on the Sabbath." *De Goddelijke Openbaring in de eersts drie hoofdstukken van Genesis*. 1932, p 30.

Here we refer to the mythological narratives of Babylon, Sumeria, Egypt and numerous other ancient civilisations, which were made accessible by archaeological research. In this connection one can think in particular of the old Babylonian *Enuma elisj*, which are the first words of a poem or song dealing with creation. Then there is the so-called *Shabaka Stone* from Egypt. This explains how creation was brought about by the word of the god Ptah, whereafter he rested.[2]

Although the lucid biblical portrayal of creation does not contain any of the typical mythological elements, nevertheless, there are similarities between the biblical narrative and these other stories. Both the biblical and the other narratives contain the same style of parallelism and the repetition of phrases. As regards content, the old mythological stories reflect some similarities with the Genesis account. The following are significant: the world and people emanating from physical deities; the division of elements in creation; the image of God after which man was created; God's rest after his work of creation; the Fall, and so on.

On the grounds of these similarities, some scholars deduced that the roots of the creation narrative of the Bible should be sought within these traditions. Thus the biblical creation narrative and other such narratives are more or less regarded as equal. Those who confess the divine inspiration of holy Scripture believe that, even if the author/s of Genesis made use of contemporary tradition about the origination of all matters, the Holy Spirit enabled the author/s to distinguish that which was true from false data. Moreover, matters were revealed to the author/s that would otherwise not have been known.

The measure of similarity of the Genesis record with the old mythological narratives probably indicates that the first people had a more or less accurate knowledge of creation. Since the art of writing did not exist at the beginning, such knowledge was transmitted orally from one generation to the next. The fact that this transmission could have taken place over a lengthy period, especially together with the sinfulness and superstition of man, could have resulted in the distortion of the facts about the creation story. These narratives became so fairy tale-like and surrealistic that they cannot be placed on a par with those in the Bible, but should be classified as of a different nature.[3] This is clearly seen when one studies the various mythological narratives. Therefore a specific divine revelation and inspiration was necessary to enable the Genesis author/s to produce the true creation narrative.

1.2 The creation narrative of the Bible as a prophetic-historical communication

A critical reading of the first chapters of Genesis raises the matter of literary genre. Is it history the way one knows it; or is it symbolic reporting; is it a poetic version of events; or is it merely a myth without any factual basis for its content?

Primarily we should keep in mind that the Bible is divinely inspired, and for that reason it does not quite fit any literary genre. Although the portrayal of the creation events has been written in a particularly narrative and his-

2 Cf M A Beek, *Aan Babylons Stromen*. 1951, Chapters iv–vi.
3 Cf C Leupold, *Exposition of Genesis, Volume One*. 1990, p 30.

torical style, it is nevertheless also presented in a prophetic-historical style. Isaiah 45:12 is an example of this. In this passage of Scripture the Lord declares the following through his prophet: "I have made the earth, And created man on it. I stretched out the heavens, and all their host I have commanded."

The creation narrative contained in Genesis 1 and 2 is not ordinary history. It was not construed from data which was merely obtained from historical research and then arranged into a logical and chronological system. Here we have to do with pre-historical and pre-scientific facts and content that could only have been obtained by divine revelation. There had to be a specific working of the Holy Spirit in the authors of the Bible. They had to obtain by revelation, that which they would otherwise not have known. They would have had to discern between the true and the false concerning the information available to them or which was in circulation at that time. Both the revealed as well as the already existing information had to be compiled into one whole and formulated in such a way that it could serve as God's Word to mankind. It had to be presented in such a way that, while on the one hand, it interpreted God's revelation and inspiration, on the other hand, it was comprehensible to the readers of that time, and to all readers thereafter. Thus we maintain that the biblical communication in connection with creation is presented in a prophetic-historical way.

This prophetic-historical narration is interpreted and proclaimed (kerugmatic) history. It is called interpreted history to indicate that it does not merely refer to facts from the distant past. These facts have been so interpreted that they have a deeper meaning; they speak of spiritual and supratemporal realities. We see this in the victory of light over darkness; the replacement of conditions on earth which were hostile to humankind with conditions that were friendly; the slumber into which Adam fell and the woman who was made from his body; the Garden of Eden; the fellowship that the first human couple enjoyed with God; the tree of life; the tree of knowledge of good and evil; the seduction of man by the snake; the pronouncement of judgement over man and the snake; the promise made to the woman as mother; the clothes of skin made for man by God, and so on.

By kerugmatic history is to be understood not only the communication of historical important facts to humankind. It also implies that by means of this communication God is addressing man in order that he will acknowledge and honour Him as Creator, that he will turn to God in humility and dependence, and that he will understand God's purpose with him as mankind

When reference is made to the prophetic element in the historical communication, we should keep in mind that the historical books of the Bible were written by prophets in prophetic language (1 Chr 29:29; 2 Chr 9:29; 12:15; 13:22; 20:34).[4]

4 About the fact that the prophets wrote the historical books, H Bavinck comments as follows: "The prophets not only repeatedly referred in their addresses and writings to the history of Israel, but they were also the ones who kept, revised and delivered it to us. Their intention with this was, however, not to present us with precise chronological narrative about the experiences of the nation of Israel, as other historians would deal with the subject. The prophets' approach of the historical books of the Old Testament is based on that of the torah, and from that viewpoint they viewed and described the history of Israel. Judges 2:6–3:6, 2 Kings 17:7–23, 34–41. The historical books are a commentary in facts of God's covenant with Israel." (translation – F.M.P.) *Gereformeerde Dogmatiek*, I. pp 361–362.

1.3 The creation narrative and the central message of the Bible

As with any other parts of the Bible, the creation narrative is also presented as proclamation (kerugma). Its message would finally receive its actual content in Christ and his gospel.

The main idea and intention of Holy Scripture is to relay God's word to mankind. The creation narrative, accordingly, does not present all the relevant details which one usually expects from historiography. It immediately introduces us to the essential matter, namely God's history with man. The creation narrative may be compared to a stage adorned by the minimum of decor and detail in the form of historical data. By means of the main historical facts, God's words and deeds as intended for mankind are presented. (This historical basis totally distinguishes the biblical creation narrative from that of the pagan traditions and legends. The latter are primarily concerned with the struggle of primeval forces and gods outside of any historical framework.)

In the first place, the Bible is concerned with the reporting of those things which, somehow, are related to Jesus Christ and his gospel. As a result only selected historical facts are reported. Moreover, as a result of this kind of reporting, the selected facts are not always presented in a logical and chronological order.[5] This free application of historical events with a view to the central message, contributes to a more powerful accentuation of this message.[6]

Precisely because it concerns the prophetic-historical narrative of the Bible's central message, it implies that the historical facts are not encapsulated in the past as is usually the case with ordinary history. The meaning and impact of the historical facts supersede the past and also involve the future. In this way a unity is brought about between the past, the present and the future. This is because protology, together with eschatology, has its meaning in Christ, who is the Alpha and the Omega. Therefore, God's revelation through Jesus Christ is indeed the key to the understanding of the entire Scripture, from Genesis to Revelation. Viewed in terms of the biblical doctrine of creation, He is the One from whom, by whom and unto whom all things are (Rom 11:36; Col 1:15–17).

1.4 The creation narrative as inspired saga

Karl Barth places the particular nature of the creation narrative within a different framework. He postulates that the creation narrative is unhistorical, or more accurately, that it is prehistorical history. Thus he calls it a saga, although he means something different by creation saga *(Schoupfungssage)* than what is usually understood by a saga. He qualifies this creation saga further by calling it divinatory and poetic saga. *Divinatory* is understood to mean that view or notion that one has (or receives) about that which is hidden, obscure and impenetrable to ordinary research, yet in which history has its roots. *Poetic* refers to the literary style used to present

5 J A Heyns puts it thus: "To summarise: the history which the Bible writers ultimately present as the result of their special way of work, is not the historical kind of history, but prophetic history, which by nature is episodic, fragmentary and not causal continuance, demonstrative and not merely descriptive, multi-dimensional and not only single dimensional." (translation – F P M) *Brug tussen God en mens*, p 121.

6 Cf W Eichrodt, *Theologie des Alten Testament.* 1961, 1, p 113 ff.

these prehistoric truths to us.

Barth also asserts that Genesis 1–3 is presented in the form of a narrative saga which, together with the historical reporting, is part of a particular proclamation. We find – and this Barth says with caution – very little pure history and also very little pure saga in this proclamation. This does certainly not imply that we are dealing with mere speculation and flights of fantasy; it is not like a myth. (A myth is not based on any real events, and usually deals with gods and legendary figures who never existed). Although the creation narratives found in Genesis are prehistorical, they are nevertheless based on real events. The intention of the Bible is thus that it should be understood literally.[7]

1.5 Primeval History and the Beginning of History

On the grounds that the biblical creation narrative was compiled from old mythological writings and narratives, Westermann distinguishes between primeval history *(Urgeschichte)* and the beginning of history *(Anfang der Geschichte)*. Primeval history is that which deals with creation and the first people, and is referred to by the first chapters of Genesis and the mythological communications. This primeval history is completely inaccessible to scientific study. However, it is essential as background to the actual history of God with humanity as it originated at approximately the time of Abraham. Thus, no actual historical line connects the creation events to ordinary human history. The genealogy of Adam to Abraham would have been added by the author(s) of Genesis in order to connect primeval history with the beginning of the history of Israel.[8]

1.6 The creation narrative in parallel versions

The creation narrative differs from ordinary historiography, because it deals with so-called doublets, that is, the same narrative is presented in two parallel communications. Thus, we have one version of creation in Genesis 1 and another in Genesis 2. Theologians distinguish them by attributing the first to the Elohistic and the second to a Jahwistic authorship. (This is because Genesis 1 speaks of God (Elohim), whilst Genesis 2 refers to Lord (Jahweh)).[9] The account of Genesis 1 takes us through the principal moments of creation, until it reaches its climax and completion in the creation

7 *Kirchliche Dogmatik* 111, 1, pp 87–92.
8 Claus Westermann, *Biblischer Kommentar Altes Testamentes*, pp 58, 111–116, 227, 228.
9 A view exists that the Pentateuch (the first five books of the Old Testament) was compiled from four documents. They are the Elohistic, Jahvistic, Priestly and Deutronomic, which are respectively indicated with the letters E, J, P and D. These primeval documents would be distinguishable from one another on the grounds of difference in language, style, vocabulary, content etc as appear in parts of the Pentateuch. These documents were supposed to be summarised into a unity by an editorship (R) which took place more or less in the time of the exile of Israel and thus as late as 400 years *bc*. In connection with the above statement we would like to mention that Christ and the authors of the New Testament associate their quotations from the Pentateuch with Moses as author. Matthew 8:4; 19:7, 8; 23:2; Mark 7:10; 12:26; Luke 16:29, 31; John 7:19, 22, 23; Acts 3:22; Romans 10:5; Hebrews 9:19.

After pointing out the subjective and doubtful grounds upon which Bible critics base their statements, H C Leupold mentions *inter alia*: "the beautiful and consistent harmony of the individual accounts, a harmony which is in itself the strongest index of single rather than a composite authorship". *Exposition of Genesis*, Vol. 1, p 20. The fact that the death of Moses is recorded in Deuteronomy 24 indicates that other people must have supplied additions. What concerns us is that Moses, together with whomever contributed to the writing of the Pentateuch, wrote under the inspiration of the Holy Spirit.

of humankind. The second account (Gen 2:4–25), only mentions that the Lord God made heaven and earth, and thereafter elaborates more extensively on the creation of the first man and his wife. Subsequently, it presents the garden in which the Lord placed the first human pair and the events that took place there.

What we read in Genesis 2 should, however, not be seen as an alternative creation narrative to that of Genesis 1. Instead it is a case where the account of Genesis 2 links up with that of Genesis 1 in order to provide more detail about man and his earlier history.

These kind of doublets frequently occur in the Bible. In the Old Testament we find that much of the history contained in the books of Kings is repeated, although somewhat differently, in the two books of Chronicles. We find something similar in the four Gospels of the New Testament.The generally accepted principle in interpretation is to merge these dual accounts and reconcile them with each other.

2 A CLOSER LOOK AT THE CREATION NARRATIVE

In Genesis 1 we read: "In the beginning God created the heavens and the earth". These words infinitely supersede anything that can be said or imagined by man. It explains why there is not just *nothing* and *why* everything *is there,* including man himself. If God did not create, there would have been nothing that could have been observed. Whatever man may say or do today; whatever he sees in himself and his world; whatever he may discover in the universe (macroscopically as well as microscopically); whatever he may deduce from his observations, all these matters would not even have been relevant if God did not first create everything.

2.1 The meaning of Genesis 1:1

We now give greater attention to what is presented in the first verse of the Bible.

2.1.1 The meaning of: In the beginning

The words *"in the beginning"* (Heb: *berègshit*) presupposes a specific moment in time, looking from the present to the distant past, when the visible cosmos (heaven and earth) was created. Together with this, we have to say that time, like space, is part of all things which can be observed or conceptualised. We who ourselves have a temporal-spatial existence, can in no way observe or imagine anything that did not exist at a certain time, at a certain place or in a certain space. Therefore, time has to be seen as something which originated together with everything else that has been created. Thus, time is something that also had to have been created.[10]

If we were to ask what was prior to that beginning when everything was created, we dare not think of any time that preceded it, for the simple reason that there was no time before that. That which was there before time, or better said, that which encompasses all time, cannot be something that lies on the level or continuum of time. Here we are referring to eternity, (some-

10 Saint Augustine already remarked that the creation did not take place in time, but together with time.

thing which one cannot conceptualise) and which is related to God himself. We pointed out in a previous chapter,[11] that there is no way in which one can imagine what God is like, and in that way objectivise him spatio-temporally. Man as subject can at most know that he is from God and that he has to do with God, since God has revealed himself to us because of his grace.

The reference in Genesis 1:1 to *the beginning*, is made with regard to the moment when all things became real in their temporal-spatial existence. The expression *heaven and earth* should be understood as referring to the entire universe.

2.1.2 The meaning of: God created

This is firstly an acknowledgement that God *is there,* and that He created everything. There are two options regarding the origin of the universe. It either is a case of believing that God created the universe, or believing that the stars and the earth and everything on it were able to bring forth themselves out of nothing. It is obvious that the latter view is a complete contradiction. The latter also presupposes the possibility that matter is able to develop by itself into self-conscious and highly intelligent beings (humans) with a will of their own. Thus, it is a case of either declaring matter to be God (or at least ascribing capabilities to it that can only be associated with God); or a case where one should believe in the eternal Creator God, who himself is exalted above matter with its spatio-temporal limitations.

It is not on account of logical deductions or scientific findings that we confess that God created everything, but on account of his revelation of Himself to us. This self-revelation of God to us through his creation can only be understood by faith, that is, in a personal relationship with God. For this reason we confess together with the author to the Hebrews: "By faith we understand that the worlds were framed by the word of God, so that the things which are seen, were not made of things which are visible." (The NIV puts it this way: "By faith we understand that the universe was formed at God's command, so that what is seen was not made out of what was visible" Heb 11:3.) The fact that God created everything is an integral and inseparable part of both the Old and the New Testaments (cf Isa 41:20; 45:18; Amos 4:13; Mark 13:19; 1 Cor 11:9; Eph 3:9; Rev 4:11 and 10:16). Without this confession of God's creatorship the entire biblical revelation has neither meaning nor relevance.The Bible teaches that everything was created by the Word; that is, through Jesus Christ (John 1:3; Heb 1:2). God, through Jesus Christ is the One from whom, through whom, and unto whom, all things are (Rom 11:36). Hebrews 11:3 puts it clearly that the world (and together with that the entire cosmos) did not originate from that which was there before. Theologians speak of a *creatio ex nihilo* (a creation from nothing). However, this does not imply that the *nothing* was the origin of the creation. It only implies that that which was created had no existence in itself. It is only from God Himself, that that which was not there at all, came into being. Romans 4:17 speaks of God who "calls into being what does not exist." (Moffatt Translation).

11 Cf Chapter 7, *Words of light and life. Volume 1.*

2.1.3 The meaning of: To create

The Hebrew word, *bara* (create) as it appears in Genesis 1:1 is always used of a divine act. This word is always used for the origin of everything, and therefore it can never be understood as an act of man or of nature.

In the Bible the word *bara* has a secondary meaning, namely where God brought forth something or created a condition from that which already existed (Gen 1:21, 27; Ps 102:18; Isa 54:16; 65:18; Jer 31:22). Sometimes the word *bara* is therefore used instead of *asah* (to make), as it appears in Genesis 1:26, and *jatsar* (to form) as used in Genesis 2:7.

2.1.4 The implications of the fact that God created everything

We can make certain deductions from the fact that God created everything.

Firstly, it implies that the visible universe or cosmos could not exist by itself or simply not ever have had a beginning. There had to be a moment when it became a visible reality.

The fact that everything was *created* forbids the worship, absolutisation or exaltation of anything in creation to the level of deity (for example: the sun, earth, laws of nature, natural phenomena, etc).

Furthermore, it implies the rejection of any form of ancient heathen dualism, according to which matter always existed besides God. The Manicheistic view which regards matter as evil and existing independently from God who is good, is to be rejected. With this we also reject Gnosticism that ascribes evil to matter. God not only created *everything,* but He created everything *good,* so that it may fulfil his purpose.

Moreover, it also rejects the idea that creation is an emanation from God. Such a notion implies that creation is an extension of God himself, and thus creation is also divine. This is what we call pantheism, the doctrine which teaches that all matter is God. Although God created everything and continues to sustain everything, there is nevertheless an inconceivably great distance between the Maker and that which was made; between Creator and creation.

Since God is the Creator of everything, his handiwork reflects something of his majesty, greatness, wisdom, power, etc; in fact, creation is a paean of praise to the glory of God.

The fact that God is the Creator and Sustainer of everything, implies that no-one has a final and eternal claim of possession of anything. Everything has been lent to man for a while. For this reason man should deal with creation in an attitude of respect and concern. If man were to harm God's handiwork – especially in the form of his fellowman – then he transgresses against God himself.

2.2 The interpretation of Genesis 1:2

In Genesis 1:2 one reads: "The earth was without form, and void; and darkness was on the face of the deep. And the Spirit of God was hovering over the face of the waters."

The problem of interpreting the words: "void" (Hebrew: *tohu wabohu*) has given rise to a number of different explanations. We mention some important ones.

2.2.1 Remains of ancient Babylonian cosmogony (doctrine about origin of the cosmos)

As already mentioned, there are ancient heathen conceptions about how the cosmos as man knows it originated. This was seen to have happened when a god or gods wrestled with primeval forces and overcame them. A material universe supposedly emerged from these primeval struggles.

Those belonging to higher-critical circles postulated that Genesis 1:2 speaks of the influence which this old Babylonian cosmogony had on the writers of the Old Testament. *Tohu wabohu* in this passage would have alluded to the existence of these ancient forces.[12]

In support of this view, it is said that in the rest of the Old Testament there is also reference to these monsters overcome by God. Reference is then made to the heads of the sea monsters and the Leviathan that were crushed (Ps 74:13, 14), the coiled serpent and dragon that were killed (Isa 27:1), Rahab which was destroyed (Job 9:13; 26:12, 13; Ps 89:11), and others. However, we should pay attention to the fact that the biblical style and content is completely different to that of the ancient heathen myths. The Bible (as was previously indicated), speaks of divine inspiration.[13] Indeed the Bible does refer to the above monsters, but these references can easily be identified as poetic-symbolic language, used for Satan and his evil angels. We also find similar references to Satan in Revelation 12:3, 9; 20:2. There is therefore no indication in the Bible that *tohu wabohu*, in Genesis, is connected to these kinds of monsters.

2.2.2 Primeval matter from which the earth was made

The very first question to be asked is how Genesis 1:1–3 should be translated. Besides the usual translation, there is the following: "When God began to create the heavens and the earth, the earth was a desolate waste, with darkness covering the abyss and a tempestuous wind raging over the surface of the waters. Then God said, 'Let there be light ...' ".[14] According to this, verse 2 is placed between brackets (parenthesis), while verse 1 is made a conditional clause and verse 3 is made a principal clause.

Although it is not grammatically incorrect to read Genesis 1:1–3 in this way, it does indeed create serious problems.[15] If the first verse of Genesis 1 is not regarded as an independent idea alongside verses 2 and 3, then God's act of creation is not regarded as an act in its own right. It simply refers to a temporal reference to a time when certain conditions were prevalent on earth. If the latter were the case it would imply that Genesis 1:1 and 1:2 only constitute the setting for the words of verse 3: "Then God said, "Let there be light"; and there was light." Reference to the most important events (the creation of heaven and earth) is then made subject to

12 Cf A Noordtzij, *Korte Verklaring*; E Sauer, The king of the earth. 1962, pp 197.

13 Cf Chapter 5, *Words of light and life. Volume 1.*

14 Cf Meek, *An American translation*, as quoted by Wilbert H Rush in Darwin, Evolution and creation. 1972, p 48.

15 This corresponds with the translation of Leidsch. Cf G Ch Aalders, op cit, pp 195–196.

something of far lesser importance. Such a reading of the first three verses of Genesis thus undermines the entire idea of creation. It is no longer said that God created everything, yes even the very matter from which the earth was created, but rather that He worked with already existing matter. Although reference is made to God's work of creation, it does not mean the bringing forth of that which existed before, but merely the orderly arrangement of that which was already there and of which the condition is described as void and empty.

2.2.3 Genesis 1:2 and the doctrine of restoration (restitution)

This view propounds that Genesis 1:1 speaks of the perfect creation of the universe, whilst verse 2 speaks of a condition into which the earth fell on account of the fall of Lucifer and his angels. This presupposes that Lucifer had rulership over the earth. At his rebellion God did not only judge him, but also the whole earth; hence we read that the earth was without form and empty, and that there was darkness on the surface of the deep.

Because of God's intention of placing man on the earth, this chaotic condition had to be restored to order. Genesis 1:3–25 is about the restoration of the earth to the perfection which it had in the beginning. Thus, what took place during the six days of creation is a restoration of the earth.

In support of this view, it is propounded that creation is a self-revelation of God, and for this reason everything He made is good and perfect. Therefore the condition described in Genesis 1:2 which speaks of chaos cannot be the condition of the original earth.

To support this view further, Isaiah 45:18 is quoted: "For thus says the Lord, who created the heavens, who is God, who formed the earth and made it, who has established it, who did not create (Hebrew: *bara*) it in vain (Hebrew: *tohu*), who formed it to be inhabited." Reference is also made to Job 38:6–7 where it is mentioned that the morning stars and all the sons of God rejoiced together when God created the earth. These beings were most probably angels, and they could not have rejoiced unless the earth was good and perfect.

The fall of Lucifer and his angels, as well as the concomitant judgement of the earth over which they had jurisdiction, must have taken place before the creation of living beings and of man. The best place to fit the outcome of this judgement over the earth would be in Genesis 1:2.

It is further argued that the combination *tohu wabohu* as it appears in Genesis 1:1, also occurs in the rest of the Old Testament, for example in Isaiah 34:11 and Jeremiah 4:23–26. In the later passages it speaks of the effects of God's judgement. Therefore it would be correct to understand the condition of *without form and void* as it appears in Genesis 1:2 to be a result of divine judgement.

It is also argued that *bara*, (Hebrew: create) used in Genesis 1:1, is a reference to the original creation, whilst the *asah* (Hebrew: make or form) is used in the rest of Genesis 1 to refer to the origination of man and the animals. The use of these two different Hebrew words should be ascribed to the difference of the original creation and the restoration of the creation that came under God's judgement.

Furthermore, notice that traces of this doctrine can be dated as far back as Augustine (approx 400 AD), and have since been propagated by numerous prominent personalities.[16]

2.2.4 Critique of the doctrine of restoration

Much argument could be brought to bear against this viewpoint. The most serious is that the biblical narrative shows no record of a creation and a re-creation in the earth's history; there is not even a hint of something like this. After all one would expect that such momentous and incisive events would have been clearly mentioned, especially when we think of how extensively, infinitely less important matters are treated in Genesis 1.

It can hardly be accepted that the Bible would only devote a single brief statement (Gen 1:1) to this original and perfect creation, whilst the entire rest of that chapter is devoted to the so-called restored creation. In any case, it is very difficult to accept that the moon and the stars were destroyed by judgement and then had to be recreated. However, it is true that Lucifer and his angels could have had a detrimental effect on the earth, and that this also could have lead to the fall of man. On the other hand, the Bible does not link the fall of the angels to the conditions on earth as described in Genesis 1:2.

Isaiah 45:18 does not refer to a judgement which struck the earth, but of a development which the earth underwent, so that it would, for man's sake, not be empty and void, but habitable. The emphasis on *inhabit* as the purpose of God with the creation of the earth, is a confirmation of this interpretation. In this verse no distinction is made between a universe that was created and one that was re-created.

Furthermore, we should take note that the strict distinction drawn between *bara* (create) on the one hand, and *yatsar* (form) and *asah* (make) on the other, cannot be justified on the basis of the Bible. We have already seen that *bara* can also have a secondary meaning and may also be used in the same sense as *yatsar* and *asah*.

The poetical-symbolic quotation from Job 38:7 where the angels, who rejoiced over the foundation of the earth are mentioned, gives no indication that this reference has anything to do with a former perfect earth. One should deduce from the context that the earth, over which the angels rejoiced, was the same earth with which Job was familiar. There is no reason to surmise that there is reference to the earth in a former state.

Moreover, the notion that Genesis cannot refer to the self-revelation of God since it presents chaotic conditions, cannot be maintained. At best we can deduce from this verse that the earth was still uninhabitable and impossible for man to live on.[17] Although uninhabitable, that which God created still reflects his glory, particularly when we behold the stellar heavens.

Furthermore, a previously perfect earth with the possibility of living beings also presupposes that there were in fact living beings on it. If that earth was destroyed by God's judgement, then all living beings would also have died and been consigned to a watery grave, that is, the flood which

16 Cf E Sauer, op cit, pp 195, 196.

17 With reference to this, Leupold declares: "Now *tohu* as such means originally unformedness (K. W.) and so can come to mean a 'waste' only in the sense of being not yet put into shape." Op cit, p 46.

covered the earth. Geological studies indicate that there are no geological strata which provide evidence of destruction and which distinguish a previous earth from the earth in its present condition.

Palaeontological research shows that there are no two groups of fossils of living creatures which have been separated for a long time from one another. Rather it proves the opposite, namely that all fossils of prehistoric creatures form part of one system.[18]

We should, furthermore, keep in mind that the Book of Genesis was not only intended for a number of well-informed theologians who would be able to deduce the doctrine of restoration from Genesis 1:2. The authors as well as the original readers were ordinary people. In terms of our civilisation they were actually on a very elementary scientific level. Therefore the presentation of Genesis 1:2 had to be such that the readers of that time would have been able to understand it.

2.2.5 Genesis 1:2 and Satanic influence

After his discussion of *tohu wabohu,* S du Toit concludes that this phrase does not merely indicate a neutral position of the earth during its process of development. He points out that *tohu* occurs 19 times in the Old Testament. Eleven of these occurrences which are found in Isaiah have the basic meaning: emptiness, futility, insignificance, nothing. It is also used to refer to the terrifying appearance of the desert (Deut 32:10; Ps 107:40); the insignificance of the idols (1 Sam 12:21); destruction after God's judgement (Isa 24:10; 40:23, 24). Related to *tohu* it describes the scene after God's judgement struck a nation or a region (Isa 34:11; Jer 4:23). When we combine this with the darkness and the earthly flood (*tehoom*; or the deep) as is mentioned in Genesis 1:2, then we have a picture of something ominous and frightful to mankind (but, in particular, to the Israelite). This state is identified as the outcome of the work of Satan and his followers. The power of evil is thus seen as something active right from Genesis 1:2, until the end of all things (as depicted in the Book of Revelation) when the primeval serpent and the dragon will be finally destroyed. Although these forces were at work from the very creation of the world, every act of creation checked the working of the evil forces. Thus, there is no idea here of a struggle *at* the creation (as in the heathen cosmogenies), but of a battle *after* creation. Although *tohu wabohu* speaks of Satanic influence, there is nevertheless, no mention of a perfect primeval creation that was destroyed and then again restored, as propounded by the restoration theory.[19]

In response to this view, one may reiterate that the Bible nowhere links the *tohu wabohu* of Genesis to Satan. As already pointed out in connection with the doctrine of restoration, there could be conditions so inhospitable to man that they evoke fear and anxiety. This could be on account of divine judgement which struck such a region; but it could also be that such an area was not (or is still not) intended for man to dwell. This would be the case with regard to the condition on other planets besides the earth in our solar system – not to mention the sun and the other stars! When conditions on

18 CF E Sauer, op cit, pp 213, 237–240, 250, 251.
19 S du Toit, *Bybel, Skepping, Evolusie*, 1968, pp 164–166, 172–176, 339.

those planets are described by modern science, then the very idea of an unprotected person ending up on them fills one with fear and trepidation. And yet, that does not imply that those conditions were caused by evil spirits. Thus one has no right without further ado, to maintain that the condition of the earth, as described in Genesis 1:2, was the work of Satan and his followers.

2.2.6 Genesis 1:2 speaks of the dark aspect of creation

Karl Barth acknowledges the difficulty (crux interpretum) of interpreting Genesis 1:2, and points out that in the time of Augustine there were already no less than five different interpretations.[20] Barth sees the conditions referred to by the text as something that was not brought about by the word of God, and therefore had no right of existence and continued existence. He views this as the *nothing* or the chaos that only has meaning because it is the antithesis of what God intended. It is something which has been denied, rejected, passed by and left behind. Barth explains that the purpose of God's creative work was a universe, which had to exist as opposed to this nothingness or chaos.

Furthermore he says that this creation has a positive and a negative, a light and a dark side. This negative or dark side is opposition to, and forms the background of, everything that is good.

This *anti-creation* side, however, is a force which evidences itself in creation in the form of disasters, mishaps, suffering and death. This can be compared to sin, which Barth elsewhere describes as the impossible possible.[21] This dark side of creation, this anti-creation force exists for the sole purpose of confirming God's victory over it. Together with this Barth also says that creation with both its positive and negative aspects exists with a view to the re-creation. This re-creation is the redemption and salvation brought about by Christ. In brief, it amounts to creation actually existing with a view to the re-creation through Christ.

However, we maintain that it was not the intention of the author of Genesis to convey such philosophical and metaphysical ideas to his readers. Moreover, by this reasoning one gets caught in all sorts of dualism and contradictions. This can be seen from the inference that God built both positive and negative influences into creation; that everything was created in order to be re-created; that there is darkness that light might triumph over it.

It should be noticed that this verse says: "And the Spirit of God was hovering over the face of the waters".[22] From this it should be deduced that we are not concerned with *nothing*, or something which God has rejected. The creative Spirit of God had a particular relationship with it. It should rather be

20 *Kirchliche Dogmatiek*, 111, 1, pp 112–120.
21 *Kirchliche Dogmatiek*, 111, 3, p 403 ff.
22 *The new English Bible*. (Oxford Press) 1970, translates the Hebrew word *ruach* with "wind" and renders the second half of Genesis to read as follows: "... and a mighty wind that swept over the surface of the water." In this instance the expression "of God" is used as a superlative description denoting mere "mighty" Such a view is, however, strange in the context where it often speaks of God as Creator. The word "hovering" (Hebrew *meragefet*) does not signify something static but something active. This word also appears in Deuteronomy 32:11 where mention is made of an eagle "hovering over its young". This excludes the term "brood" as an alternative translation.

seen as an indication that the Lord concerned Himself with the earth in that state, and was not turning his back on it.[23]

2.2.7 Genesis 1:2 is a logical connection between Genesis 1:1 and the rest of the chapter

It should be accepted that Genesis 1 was written simply and clearly so that any reader of that time and thereafter could understand it. Genesis 1:2 should therefore have an inevitable and a logical place within the creation narrative. It was pointed out in the above that the purpose of the creation narrative was to provide the background and a run-up to God's history with man. In Genesis 1:1, one finds the general fundamental statement that in the beginning God made heaven and earth. After this the creation of earth is singled out, and greater detail is supplied as to how it originated and came to be as we know it.

We can deduce that the creation of the earth in a way that it could sustain life, did not take place in one moment. There was a stage, as described in Genesis 1:2, where it would have been impossible for any life to dwell on it. However, during the process of creation, the earth and its conditions were changed so that living creatures could be created and could find a home on it. This process would not have been necessary and would have been meaningless if it did not take place, as described, against the background of Genesis 1:2. After the general introductory statement of Genesis 1:1, Genesis 1:2 indicates the commencement of the process whereby God prepared the earth for living creatures and finally for man himself.

2.3 The days of creation

From Genesis 1:3 we read how the Lord, during a process lasting six days, prepared the earth for living creatures to appear on it, and finally for man himself. However, there is a sharp difference of opinion about the understanding of these six days, also called the *hexahemeron* (with reference to the Greek for *six* and *day*).

2.3.1 Do the six days indicate the order of creation?

There are those who maintain that creation did not take place over six distinctive days. The division of creation into six days only speaks of the projection of the author who wished to link creation to ordinary biblical history.[24]

There are also those who maintain that is not a matter of a real division of time into days or periods. The creation narrative is not concerned with a process of events which ran according to a particular natural-historical

23 With reference to the Spirit of God who was hovering over the face of the waters, Leupold maintains the following: "We could never believe that this hovering of the Spirit over the face of the waters was idle and purposeless. From all other activities that are elsewhere ascribed to the Holy Spirit we conclude that his work in this case must have been anticipatory of the creative work that followed, a kind of impregnation with divine potentialities, the germs of all that is creted were placed into dead matter by Him." Op cit, p 50.

24 Cf C Westermann, op cit, pp 125, 126.

order, but it is about an ideal order as found elsewhere in the Bible.[25] God's work is seen to be so unfathomable that it cannot simply be confined to a time scheme or a logical order.

Contrary to these viewpoints, the following can be pointed out. If one does not consider the days, as presented in Genesis 1, as real time divisions, then one has a problem with the events which took place on those days. If an ordinary reader does not consider it to be real time periods which follow on one another, how can reality be ascribed to the events on those days?

The remark following each day: "And God saw that it was good" would then also have no sense.The fact that there is a different kind of order of things mentioned elsewhere in the Bible, in no way denies the existence of such temporal order. In no way can the difference in any other order be used as basis for denial of a succession of real days or periods in Genesis 1!

Furthermore, it should be noticed that the law for the Sabbath (Exod 20:8–11) is based on the fact that God made heaven and earth in six days and rested on the seventh day. If one does not accept this time division and temporal order in Genesis 1, then one has a problem with the reason why the Sabbath was instituted. The events surrounding the creation of the earth is not something which God has hidden in unfathomable things. It is something to which He attached a time dimension, so that it could be described in terms of time and temporal order.

2.3.2 Length of the days of creation

There are those who maintain that the days of creation extended over 24 hours, just as we know days. They base this on the usual meaning of the Hebrew word for day, *joom.* They point out that it is a sound hermeneutical principle to understand any word in the Bible in its most ordinary literal sense. It is further inferred that the reference to *evening* and *morning* is an indication of days as we know them. Furthermore, they explain that the law of the Sabbath, based on the days of an ordinary week, can actually be traced to the creation week. It is also maintained that if the author wanted to refer to periods of millions of years, rather than 24 hour days, he would have done so. Also the expression "and it was so" which follows the creative word spoken by God, is interpreted as referring to the rapid succession of the creation events – so quickly did it take place that it could fit into 24 hours. It is also pointed out that whenever a number is related to a

25 Noordtzij points out the discrepancy in the sequence of the ten plagues as recorded in Psalms 78 and Exodus 7, the temptations of the Lord as mentioned in Matthew 4 and Luke 4 as well as the facts about the life of Hezekiah narrated in 2 Kings 18–22 and that which we find in 2 Chronicles. On the grounds of this he declares that the time division in the form of days as we read in Genesis 1 is only a projection of the author. This projection does not concern any time division or time sequence as we know it, but is simply a way in which holy Scriptures present us with the glory of creation in accordance with God's great plan of salvation. Cf Ch Aalders, op cit, p 232 ff.

day(s) (eg Gen 7:11; 8:14; 17:12; Exod 12:6), it indicates calendar days.[26]

However, this view which maintains that the days of creation were 24 hour days, is disputed by certain truths. Firstly, one should note the influence the sun exerts on the earth. The 24 hour day cycles could thus only have began from the fourth day onwards. These natural calendar days which are determined by the relationship of the earth to the sun, should therefore be seen *within* a day of creation and not *as* a day of creation. The cases referred to in the Bible where numbers are linked to 24 hour days, refers to calendar days and not days of creation. These calender days originated on account of creation. As mentioned, time did *not* exist *before* God created, but came into existence at creation as part thereof.

The Hebrew *joom* has more than one meaning. This can already be seen in the creation narrative. In Genesis 1:5 the light that was created by God on the first day is called *day*; in the following verses *day* is indicated as a period which was demarcated by evening and morning; in Genesis 1:14–16 a *day* is indicated as the period when the sun shines on the earth (something which differs in length, depending on where you may find yourself on earth); in Genesis 2:4 *day* means the entire time within which the Lord God made earth and heaven.

Furthermore, the Bible speaks of the *day* when God will visit nations with his judgement (Isa 13:6, 9; Joel 1:15; 2:1, 11; Amos 5:18, 20); the final judgement (Ezek 30:3; Mal 4:5); a particular stage in one's life (Ps 50:15; Eccles 11:9; 12:1); the lifespan of man (Ps 37:18; 103:15); period of time (Ps 77:6); opportunity in life (Ps 90:12); time of salvation (Rom 13:12; 2 Cor 6:2); the moment of Christ's return (Matt 24:36; 2 Peter 3:12) and also other meanings. In none of these passages is reference made to a 24 hour day. We have to understand the meaning of the word *day* in its context.

The argument that God made everything quickly because of the phrase "and it was so" is also invalid. This expression does not indicate the length of the days of creation. If we reason that God in his omnipotence accelerated the natural processes, geological, biological, et cetera, so that these events could take place within 24 hours, then we may as well ask why God in his omnipotence did not simply create everything in a split second. The Bible expressly teaches us that God preferred to complete his work of creation by means of successive natural processes, and that a certain time was allocated to each process. Even today, we notice processes in nature which stretch over millions of years. As an example we can mention geological phenomena. Geological data indicates that one has to think of the days of creation in terms of millions of years. That does not imply that one

26 Cf H C Leupold, op cit, pp 64–69; Wilbert H Rusch, op cit, pp 59–61. E Sauer mentions the following: "Almost all the Church Fathers, the medieval schoolmen and older Protestant theologians held that the days of the Mosaic account of creation were literal days of 24 hours. So far as Britain is concerned, special mention should be made of the seventeenth century Irish archbishop James Usscher, whose Old Testament chronology became most influential and is still to be found in most renditions of the Authorised Version. His contemporary, Dr John Lightfoot, Professor of Hebrew and Vice-Chancellor of Cambridge University, believed that the week of creation was the week of October 18–24 in the year 4004 BC and that the creation of Adam took place on the Friday of that week, October 23, 4004, at 9:00 a.m. forty-fifth meridian time. Op cit, p 193.

has to turn to science such as geology to interpret the Bible. Science can, however, question or confirm certain interpretations of biblical pronouncements.

One should always remember that the author of Genesis uses the word "days" precisely because at his time and in his circumstances there was no more suitable word to indicate the epochs of creation. The word *day* with its great range of possibilities of meaning, is the most suitable to reflect these great truths in a way that readers of all times will understand them. The word *day* can also aptly refer to the notion of *week* consisting of six days of creation, followed by a seventh day of rest. No one can determine the length of each day of creation. What we can say is that these days were infinitely longer than 24 hours. At the same time, a creation day expresses a particular limited period of time. This can be deduced from the evening and morning of each *day*, and also from the fact that the one day is followed by the next.[27]

3 THE AGE OF CREATION

Determining the age of the creation is inevitably dependent on the interpretation of the days of creation. If one maintains that these were 24 hour days, it implies that the events described in Genesis 1:3–27 took place in the duration of a week as we know it. If one could determine when Adam was created, then one would also know when everything else was made. According to this, certain people maintain that the earth and everything on it, came into existence approximately 4000 BC.

There are usually two ways used to calculate the time of Adam's creation. The first method is to calculate it on the basis of the genealogies found in Genesis 11:10–27; Matthew 1:1–17 and in Luke 3:23–38. Using this method, the number of generations from Adam to the birth of Christ are determined. This period of time can then be an indication of when Adam and the entire universe were created.

The other method of calculation is that used by dispensationalists. The proponents of this view work with the epochs (dispensations) according to which the history of the earth is said to have elapsed. They maintain that the period that man has been on earth corresponds to the week of creation. As creation took place over a period of six days and God rested on the seventh day, the same can be said with regard to the period which man has inhabited the earth. Where the Lord speaks of a *day* it has to be seen with reference to 2 Peter 3:8, as a period of a thousand years. The seventh day of human history on earth which corresponds to the seventh day (the Sabbath) of the week of creation, would be the millennium or the last thousand years of this earth (Rev 20:1–9). If the seventh day is one thousand years (a millennium) then the previous six should also be one thousand years each. The six thousand years already elapsed is divided as follows: 4 000 years from Adam to Christ, and 2 000 years from Christ to his second coming and the dawn of the millennium.[28]

27 Von Rad points out that the ancient Hebrew day consisted of day and night, and according to the cultus it began officially in the evening (Exod 12:18; Lev 23:32). *Theol Dict NT*, 2, p 943.
28 Cf Clarence Larkin, *Dispensational truth*. 1918, p 18.

One can respond to these views as follows. The genealogies found in the Bible are not meant to provide us with a scientifically based genealogy. (Our remarks on the historicity of the Genesis creation narrative are also valid here.) It is for this very reason that we find a difference between the genealogy in Matthew and that in Luke. Matthew omits a section included by Luke, in order to be able to present two lines of fourteen generations (Matt 1:17). Furthermore, there are disparate ideas as to how the expression father of and son of as found in the genealogies should be interpreted. Amongst others we should keep in mind the ancient Jewish usage of father of and son of, in which these expressions do not necessarily refer to an immediate father or son.[29] Therefore, it is actually impossible by means of this method to calculate the date when Adam was created. The way in which the dispensationalists calculate everything (basing everything on seven dispensations of one thousand years each), is mere speculation, without a biblical basis. 2 Peter 3:8 does not allow us to infer that wherever one reads of a *day* in the history of mankind, a period of a thousand years is implied. The apostle Peter is merely saying that the Lord's view of time with reference to the return of Christ is not the same as man's.

The fact that Revelation 20:1–9 speaks of a thousand years, does not authorise one to divide the entire human history into seven dispensations of a thousand years each.

We should reiterate that one cannot arrive at a date when Adam appeared on earth by way of biblical information. However, it is possible to conclude that he did not exist many thousands of years (let alone millions of years) before Christ. This is postulated because, from Adam and throughout the Old Testament, we have the same religious, cultic and cultural practices.

Even if we could have known when Adam was created, it would, in any case not help us to determine the age of the creation. It was already argued that one should not think of the days of creation as 24 hour days. One should rather think of lengthy epochs.According to astro-physical calculations, the cosmos is about 15 billion years old. This is calculated in terms of the time that it took for the heavenly bodies to move away from one another after the so-called *Big Bang* (or big explosion). This was the occasion when an overheated energy mass exploded from which all the celestial bodies began their journey in different directions into space.[30]

29 Cf A S Geyser, *The genealogical table of Jesus Christ according to Matthew 1:1–17 and Luke 3:23–38*. 1946.

30 Cf John Gibbon and Martin Rees, *Cosmic Coincidences*. 1991, p 5.

CHAPTER 2 Events in connection with creation

This chapter brings us to the events connected with the creation.

1 THAT WHICH HAPPENED ON EACH DAY OF CREATION

The work of God's creation stretches over six days. From this we can also conclude that the beginning of the week of creation is also that beginning when God created heaven and earth (Exod 20:11).

1.1 The first day

On this day God said: "Let there be light" (Gen 1:3). As with all other things, light came about by a divine fiat (Latin for the command: let this be done!) It is clear from this that God spoke and that as a result of this word, things came about. Elsewhere (Gen 1:7, 16, 25) one reads: "Thus God made ...". On account of this there are those who distinguish between the word and the actions of God.[1]

From the context we deduce that the light referred to here is the light which shone on the earth. We know that light is caused by light waves which emanate from an object which radiates energy. This source of light had to be in space from where it radiated light onto the earth. We should also accept that certain astro-physical conditions prevailed which at first prevented light from shining upon the earth. The cause for this may either have been a condition on earth, or a condition which originated with the source of light.

We read in Genesis 1:14–18 that the sun was only made on the fourth day of creation. This is when God determined that the sun would provide daylight on earth by changing night into day. (The light which shines onto the earth by night is the light of the sun which is reflected by the moon.)

Thus we conclude from the Genesis narrative that the first light which shone onto the earth did not come from the sun, as we know it. This light, however, was sufficient for the plant life which originated on the third day. This light had to come from one direction. If it shone from all directions onto the earth then the separation between light and darkness would not have

1 Westermann speaks of *Wortbericht* (word report) and *Tatbericht* (deed report), and indicates that concerning the creation of the universe, the two become one. Op cit, p 160.

been possible. Then there would have only been light. We read that God called: the light *day* and the darkness *night* (verse 5); the expanse *heaven*; the dry ground *earth* and the gathering of the water *sea* (verse 10). It can be an indication of the measure of permanence which God gave to these conditions in nature.

The fact that day alternated with night (Gen 1:5) also indicates that the earth was revolving around its axis from the beginning.

No-one knows what happened in the cosmos when God made heaven and earth. In Job 38:4–6 the Lord explains that Job (and in this Job represents the entire mankind) was not a spectator when He founded the earth. The only way that this can be known is by God's revelation. The point is that this revelation clearly mentions that God created heaven and earth. However, it is not disclosed how God did this. On the other hand, it is typical of man to inquire through his practice of science how the stars, the sun and a planet such as the earth, could have originated. Man develops theories about the things which cannot be ascertained by science. We shall now discuss a theory which may be of importance in this regard.

Reference is again made to the enormous explosion or *big bang* which would have taken place in a primeval mass of intensely high specific gravity. With this explosion gigantic fiery gas masses or nebulae were shot in all directions into space. The nebula which would later have developed into our own galaxy or Milky Way, began to disintegrate into smaller nebulae. In this way, amongst others, a smaller disc-shaped nebula, which would later become our solar system, originated. Eventually this smaller nebula also began to disintegrate. The central, and by far the largest part, lay in the middle. The parts on the periphery disintegrated as separate spheres from the central mass. On account of the gravity of the larger part of the remainder, these separate spheres began to revolve in orbits around it. Precisely because they were so small in comparison to the central mass, and because they were removed from the fiery kernel, they cooled down in space and became more solid masses. In this way the earth and other planets came into existence. In the meantime, the central fiery gas mass of the nebula also underwent changes until it later took on the form of our sun. Thus the sun, as we know it, came into existence *after* the earth. The fiery mass of gas from which the sun was made must then have been the first source which radiated light.[2]

The purpose of citing this theory is to indicate the possibility of how light could have shone on the earth, although the sun as we know it, only appeared much later in space. Regarding the creation of light, we are reminded of the prophetic-historical character of the Book of Genesis. The words: "and there was light", is not only a reference to the historical event. It is also a truth which transcends time and space; and is more than a mere historical fact. We see how this is linked to Jesus Christ the true Light who dispels darkness from the heart of man. We read about this in 2 Corinthians 4:6: "For it is the God who commanded light to shine out of darkness, who has shone in our hearts to give the light of the knowledge of the glory of God in the face of Jesus Christ." (The NIV says: "For God, who said, 'Let light

2 Cf E Sauer, op cit, pp 209, 210.

shine out of darkness,' made this light shine in our hearts to give us the light of the knowledge of the glory of God in the face of Jesus Christ.")

1.2 The second day

In Genesis 1:6 and 7 we read: "Then God said, 'Let there be a firmament in the midst of the waters, and let it divide the waters from the waters.' Thus God made the firmament, and divided the waters which were under the firmament from the waters which were above the firmament; and it was so."

The expanse or vault (Hebrew *raqia*) is space which surrounds the earth. It is called heaven in the sense that it is above the earth. As the word heaven can have different meanings, for instance air space, stellar space and the place where God dwells with His holy angels, so the expanse or vault can also have more than one meaning. It can indicate the air space or atmosphere around the earth, but it can also refer to the space in which the stars are situated (Gen 1:14, 15; Ps 19:2).[3]

The meaning of *raqia* in the context of Genesis 1:6 and 7, indicates air space or the atmosphere around the earth. It is the sphere where birds fly (Gen 1:20) and from which the rain falls (Deut 11:11).

Since air space surrounds the earth, it became possible that there could be water in the form of clouds and mist above the earth, which is different from the water on the earth.

The space or atmosphere around the earth consists of different layers of gas.[4] This does not only make rain possible, but it is also of vital importance for other reasons. We mention some of them. It contains the elements that can be assimilated and inhaled by plant and animal life. It provides a protective layer against the ultra-violet rays of the sun and against meteorites which fall to the earth. It makes winds possible and creates weather conditions on account of the heat that is reflected between earth and space, etc. All these are important for the different forms of life which would be created at a later stage.

1.3 The third day

In Genesis 1:9–10 we read: "Then God said, 'Let the waters under the heavens be gathered together into one place, and let the dry *land* appear'; and it was so. And God called the dry land Earth, and the gathering together of the waters he called Seas. And God saw that *it* was good."

According to this divine *fiat* (command), momentous changes must have taken place in the crust of the earth. Parts thereof rose and others subsided. In this way the continents and the islands came into existence while being surrounded by the seas. It is a geological fact that the earth is still unstable under its crust. This causes volcanic eruptions, the formation of islands and the disappearance beneath the sea of certain parts of land

3 The *Hebrew-English Lexicon* translates *raqia* thus: "the expanse of heaven, atmosphere, sky". 26th ed, p 251.

4 The layer of gas which surrounds the earth, reaches several hundred kilometres above sea level into the sky. The part of the atmosphere we live in is approximately eleven kilometres deep. Above that is the stratosphere which is up to twenty four kilometres above sea level and of which the upper layer is called the ozone. It is this ozone layer which protects us against dangerous cosmic rays. Above the stratosphere we have the exosphere.

which were above the water. (According to geological reconstructions, the continents first formed an entire whole. They then tore apart, on account of the movement of the plates of the earth's crust on which they rested and drifted from one another. The different oceans originated in this way. This view is in agreement with Gen 1:9–10, where the original state is described as that of a unit of waters and a unit of dry land.)

This division led to land animals and plants on earth; and, besides them, we also have the different forms of life in the sea.

On the third day the Lord created further when he said: "Let the earth bring forth grass, the herb that yields seed, and the fruit tree that yields fruit according to its kind whose seed is in itself, on the earth': and it was so" (Gen 1:11).[5]

For plants to exist, light is necessary. As already indicated, light shone upon the earth even before the creation of the sun, as we know it. It is a scientific fact that plant life is not necessarily dependent on light that has to radiate from the sun.[6]

There first had to be plant life in order to make animal life on earth possible. Plants provide oxygen while assimilating carbon dioxide from the air. With animals the converse is true, they inhale oxygen from the air, while they exhale carbon dioxide. Furthermore plants are necessary to convert inorganic material into organic material. This organic material serves as food for animal life.

1.4 The fourth day

In Genesis 1:14–18 we read how God made the sun and the moon. In our discussion of the creation of light on the first day of creation, we have already reflected on the matter of the sun which originated after the earth. It is only after the fourth day that day and night are linked to the sun. The fact that the sun and moon are mentioned together is significant. After all, the moon can only shine upon the earth by means of the sun.

We should note, however, that the sun and the moon are described in relationship to the earth. This emphasis on the earth shows that Genesis does not present a scientifically oriented account of the origins of the cosmos. Rather, it is concerned with the preparation of earth as a dwelling place for mankind. The entire creation narrative of Genesis 1 actually serves as a background and introduction to the unfolding of God's history with man on earth.

1.5 The fifth day

Genesis 1:20–22 presents God's creation of the sea animals and the birds (or flying animals) on the fifth day. This is also the first time that one reads

5 Regarding the Hebrew word which is translated grass, Leupold states the following: "the word *déshé* may well be said to include such things as mosses and other plants designed to carpet the earth." Op cit, p 67.

6 Sauer points out that a large spectrum of sunlight is of no use for plants. They only assimilate the light from red, orange and yellow rays. Thereupon he quotes the botanist, professor Dennert: "A light which only contains these rays would therefore suffice for this process. It follows that it is foolish and shortsighted to hold that the sun is indispensable for plants ... It would rather have been possible that at the beginning a more general, lightgiving, glowing mass of matter in space fulfilled the present role of the sun for plant-life." Op cit, pp 210, 211.

about *living creatures*. It should be pointed out that according to palaeontological data the sea animals originated before those on land.[7]

However, there is a difference between the sequence of the birds and the animals on land. The Bible states that the birds appeared on the fifth day and the land animals on the sixth day of creation. On the other hand, geologists and paleontologists maintain that land animals appeared before birds. (Attention will be given to the differences between the biblical presentation and that of scientific discovery. Needless to say both these sources should be taken seriously and one should always strive to reconcile the two. The Word of God as contained in Scripture can never be in conflict with the work of God's hands as we learn to know it from the visible creation.)

A possible solution to the above-mentioned problem is connected to the understanding of the Hebrew *ophicm* (singular *oph*). In Genesis 1:20–21 it is translated as *birds*. The same word is translated in Leviticus 11:20–23 and Deuteronomy 14:19 by *flying insects* or other *flying animals*. There are dictionaries that translate *ophim* as *flying creatures*.[8] If this is a more acceptable translation then the problem is solved. Scientists explain that there were in fact certain flying insects and reptiles before animals appeared on land.[9]

Along with this we should remember that paleontology is a reasonably young science and new discoveries will still be made. The possibility indeed exists that further discoveries may lead to a revision of previous conclusions.

1.6 The sixth day

On the sixth day of creation God first created the land animals and thereafter He made man.

1.6.1 The creation of the land animals

In Genesis 1:24 we read: "Then God said, 'Let the earth bring forth the living creatures according to its kind: cattle and creeping thing and beast of the earth, each according to its kind'; and it was so."

From this we should conclude that these animals, as was the case with the aforementioned living creatures, were all made from the already existing substance of the earth. How God brought this about is not explained.

The animals that were made on the sixth day are divided into three groups: the domestic animals (cattle); crawling animals (those that move on the ground, that is reptiles) and the wild animals. These animals, and in particular the mammals are the last creatures made by God before man appeared. (This order is confirmed by science.) The animals that came into existence last, show the greatest resemblance to man. This is true of their

7 Cf E Sauer, op cit, p 212.

8 Cf Brown, Driver and Briggs: *Hebrew and English Lexicon*; Koehler and Baumgarten: *Lexicon in Veteris Testament: Libras*.

9 E Sauer states the following: "In fact, winged animals existed as early as the Carboniferous period. Some of them were of astounding size, eg the flying *Titanophasma fayoli*, up to twenty inches long. Indeed, the Carboniferous forests possessed the largest flying insects in the whole history of the earth, attaining a wing span of twenty-eight inches. Thus the creation of winged animals, according to the common testimony of Genesis and geology, actually took place before the creation of animals and mammals." Op cit, p 213.

anatomical (build and limbs), physiological (function of the different organs) and neurological (the brain and the nervous system) composition.

In this regard, it is significant that the Bible explains that the plants, the sea animals and land animals were brought forth *according to their kind* (Gen 1:12, 21, 24). This implies that one species did not develop into the other, but that each had its own origin, and possibly also its own development. Since God commanded on each day of creation that the earth should bring forth different living creatures, this possibly also indicates that one species did not develop from another.

1.6.2 The creation of man

On the sixth day man was created. In Genesis 1:26 and 27 one reads: "Then God said, 'Let Us make man in Our image, according to Our likeness; let them have dominion over the fish of the sea, over the birds of the air, and over the cattle, over the earth and over every creeping thing that creeps on the earth.' So God created man in His own image; in the image of God He created him; male and female he created them."

With the appearance of man, who was created last of all, creation reached its climax and completion. From this moment onwards the Bible deals with the history of man and his relationship to God. It is a history which runs through to the end of the earthly dispensation when man will enter eternity.

Man's creation is different from that of any other living being on earth in that God first deliberated about him. This deliberation is contained in the suggestion: "Let us make man in Our image, according to Our likeness". Man was destined to be a very special and unique creature. He would be very close to God and would be a much more direct and personal revelation of his Maker.

Several explanations have been offered for the use of the plural pronoun *Us* used in the deliberation. It could be related to the Trinity; the angels surrounding the throne of the Lord (1 Kgs 22:19); pluralis majestatis (Ezra 4:18) or it could be *pluralis deliberations* (a case where God deliberates with Himself). Preference should, however, be given to the idea that the Us refers to the Trinity, since it exists before all ages. This is a case (as with some other Old Testament statements) where more is said than what the original author and readers may have had in mind. The complete and real meaning of such pronouncements is only understood in the light of the New Testament.

Notice also that man did not merely originate from the earth, as was the case with the creation of the land animals. We read specifically that man was formed from the dust of the earth, and that God "breathed into his nostrils the breath of life;" (Gen 2:7).[10] Mere natural or biological life cannot be implied in this context, for such life also pertains to animals. What is different about the creation of man is that God breathed his own Spirit (breath) into him. Thus man was made to enjoy a special relationship with God;he is an *I* created to relate to the actual *I*, namely God.

10 About *the dust of the ground*, Leupold writes: *'Aphat'* rendered dust', does not refer to dry pulverized earth only. Here, without doubt, a damp mass of the finest earth is under consideration. Luther's rendering is still unsurpassed, Erdenkloss, lit. lump of earth'. The term does not mean mud', as the sceptics irreverently declare." Op cit, p 115.

(The meaning of *image* and *likeness* will not be explained again, since this has already been done extensively in Chapter 154, Volume 1. Other aspects about the doctrine of man were also discussed in that chapter.)

It is not clear how God made man from the dust of the earth. Moreover, it cannot be said whether this happened in the same way as when God spoke the following in connection with the animals: "Let the earth bring forth the living creatures." The Book of Genesis does not aim at explaining *how* man was made, but *unto what* he was made. He was created to be the bearer of the image and likeness of God; to enjoy a personal and direct relationship with Him; to live in a personal and direct relationship with his Maker and to reign over everything upon the earth (Gen 1:26, 28).

Furthermore the creation of the first woman attests to the fact that man was made differently from the animals. First God only made man (Adam). Thereafter he made the woman (Eve) from a body part (the rib) of Adam to complement him (Gen 2:21–23). Once again, the prophetic-historical character of the creation narrative is clear. This creation of woman from man expresses a prophetic aspect, namely that the Church (as the body and bride) of Christ would be established *from* and *for* Him.

At the creation of man God blessed him and said: "Be fruitful and multiply". This corresponds to the blessing and the words pronounced after the creation of the sea animals and the birds (Gen 1:22). The difference is, however, that man was given full powers to rule over everything. Certain theologians explain this blessing as God's promise of a providential grace for all living creatures.[11]

It should be pointed out that the origination of man was not only the final step in the process of creation. It was also the climax of this process and it gives particular meaning to the purpose of creation.

After man was made and creation was completed, we read, "Then God saw everything that He had made, and indeed it was very good. So the evening and the morning were the sixth day" (Gen 1:31). The expression *very good* can be interpreted as an indication that creation served the purpose for which it was made.

1.7 On the seventh day God rested

In Genesis 2:1–3 one reads: "Thus the heavens and the earth, and all the host of them, were finished. And on the seventh day God ended His work which He had done, and He rested on the seventh day from all His work which He had done. Then God blessed the seventh day and sanctified it, because in it He rested from all His work which God had created and made."

A question which immediately arises is what is meant by the phrase *God rested from all his works.*

This obviously does not mean that God was tired and exhausted and had to recover from the exertion. God is not man, that he becomes tired and could be worn out (Isa 40:28).

The first and the most obvious understanding of this *rest on the seventh day* is that God had completed his work of creation. (This is a case of

11 Cf C Westermann, op cit, pp 221, 222.

anthropomorphism – where God or his actions are described in human terms.) In this sense rest speaks of that which has been done and has been completed. Rest (Hebrew *shabat*) as it is used in this connection literally means "to cease from the actions one has been busy with". (See eg Gen 8:2 and Isa 14:4.)

Thus, rest at the end of creation is in contrast to creation itself. Thereby the significance and greatness of creation is stressed.

However, this mention of rest on the seventh day also has a deeper significance. It is a clear example of the prophetic-historical character of the creation narrative. The following remarks explain this.

From God's point of view, the seventh day is not yet a demarcated period as were the six days of creation. This is not a case of an evening and a morning, or one day followed by the next. This day has a beginning, but it continues into eternity. The beginning (which Gen 1:1 mentions), or the beginning of the first day, was an event out of eternity. But it was also an event which, unlike eternity, is qualified in terms of time and space. God's rest on the seventh day speaks of something that has a temporal-spatial beginning, but which was then taken up into eternity.[12]

Humanly speaking, we can put it thus: In the six days of creation one is concerned with that which God did *outside* of himself. The seventh day refers to what God is *within* himself in all eternity.[13]

That which God accomplished during the first six days is, to a great extent, visible to us. Therefore it is possible to have some conception of this, however imperfect. When the Bible says that God rested from all his works, it is a notion beyond man's powers of observation. One can only try to imagine what this could mean by virtue of what the Bible reveals.

When it says that God blessed and sanctified the seventh day, it probably means his complete identification with that day. Rest indicates a state in which God finds himself. It is this state that tells of God's complete goodness, holiness, et cetera. For this reason the day that speaks of this state is blessed and sanctified.

The following also require attention. If the *rest* speaks about the state in which God is or that which essentially belongs to God, then creation speaks of that which exists from, by and unto him, but which is nevertheless separate from Him.

Attention needs to be paid to the following truths concerning the relationship between God and his creation. On the one hand, He has indeed created everything, He is sustaining that creation and is also working out his will and plan regarding it. This relationship also speaks of God's immanence with regard to his creation. On the other hand, there is an infinite distance between God and his creation. God's eternal exaltation above his creation speaks of his transcendence.

12 According to the dispensational view the seventh day, which is mentioned in Genesis 2:2, 3, covers the period from the completion of creation to the second coming of Christ. That means that we now live in the seventh day.

Straightaway, we have to point out that this seventh day refers exclusively to God. It says that God rested that day. As we shall see later on, the Bible itself explains the meaning of this rest in connection with man. It does not concern a certain period in which man lives, but it is used for his relationship with God, a certain state in which he finds himself.

13 Some theologians speak of creation as God's work to the outside *(opera ad extra)*, as opposed to his acts within the divine Trinity *(opera ad intra)*.

Looking at creation and everything we can observe in the universe, we firstly see God's handiwork. We also become aware of more than that, namely that God, the Creator, is above his creation, He is the One who is *Entirely Different* (German: *der ganz Andere*). Nothing in his creation can be compared to him.

Thus we may say that the works of the six days of creation speak of things man is capable of knowing. Conversely the divine rest of the seventh day speaks of that which supersedes creation and human thinking.

The Bible also gives a further meaning to the word rest. It is also used in the Old Testament to depict the relationship that God wants man to enjoy with him. This is seen in the institution of the seventh day as a Sabbath for man by God. This command points to the fact of the six days in which God created and then rested on the seventh day. On account of this, man was required to abstain from work on the seventh day and to sanctify (set aside) this day unto the Lord. This command is taken up in the Decalogue, and together with the other nine commandments constitute the essence of the law as it was given in its outward form to Israel (Exod 20:9–11; 31:15).

In the Old Testament the word rest also speaks of a quiet and peaceful life on earth (Deut 28:65); of Israel's experience when God gave them the promised land (Ps 95:11); of prosperity (Ps 132:13–16); of deliverance (Isa 14:3); of God's goodly provision to his people (Ps 94:13), and also of other favourable meanings.

However, it is only in the New Testament, in connection with Christ, where one learns about the actual meaning of the rest as it was promised to man. Of this true rest (Greek: *katapausis*) one reads in Hebrews 4:3–11. There one sees the rest of God after the creation; the Sabbath rest contained in the law; the rest of the promised land; *all* point to the sabbath rest of salvation which God gives us through Jesus Christ.[14] This salvation proceeds to the eternal place of rest together with Christ and God.

The seventh day, the day of the Sabbath, speaks of the completion of this creation. The first day, the day of the resurrection of Christ, speaks of the beginning of the re-creation of all things. Because the present creation has been subjected to futility and transience due to sin, a new creation is necessary. That which is linked to the Sabbath day, namely the old creation and the law, had to be replaced by that which is connected to the first day, namely the re-creation and the grace which came through Jesus Christ. This is the only way in which man could share in God's eternal Sabbath. It is a Sabbath which is not according to this creation and time, but which tells of God in his eternal state.

Thus the rest on the seventh day is not only a sign that creation has been completed but it also reflects God's purpose with creation. It contains a

14 About the use of the word *place of rest* (Greek: *katapausis*) Bauernfeind says the following: "In the NT Acts 7:49 follows Isaiah 66:1 in using it of God: 'What is the place of my rest?' Hebrew 3:11, 18; 4:1, 3, 5, 10f, on the basis of Psalm 95:11, refer to the rest (or resting-place?) of the people. As the promise of Scripture undoubtedly points beyond the servant Moses to fulfillment by the Son (Heb 3:1–6), so the rest mentioned on the very first page (Gen 2:2) points beyond Joshua (4:8) and David (4:7) to the last things. The movement from which flowed the life of creation in the first week will lead to the sacred rest of the Creator, to the seventh day. The people of God waits for this." *Theol Dict NT*, 3, p 628.

promise to man as well as to the visible creation (Rom 8:19–23). When everything has been subjected to Christ, he will, together with everything else, subject himself to God. Thus God will be all in all (1 Cor 15:27–28).

2 THE POSITION OF THE EARTH AND OF MAN IN CREATION

It has already been pointed out that the greater part of the creation narrative deals primarily with the preparation of the earth. Thereafter, it deals with the bringing forth of the various living beings on earth, until the creation of man.

2.1 The particular significance of the earth

The importance of the earth must become evident to the reader by the considerable section which the creation narrative devotes to this planet. If the earth is compared to what surrounds it in space, it is but an insignificant speck of dust. Astronomers point out that the earth is only equal to 1 297 000 part of the volume of the sun, and that the sun is but a star of average size; there are other stars which are 1 000 to 2 700 times bigger than our sun. Furthermore, the sun forms part of the Milky Way which consists of approximately 100 000 million stars, and our Milky Way (galaxy) is but one of about a thousand million similar (and even larger) galaxies.[15] The diameter of the universe is calculated at 15 000 million light years. (A light year is the distance which light travels in one year at a speed of 250 000 km per second.)

In the light of this one may ask: If the earth can be compared to a mere drop in the ocean of the stars, why is it nonetheless so important? Why does the creation narrative place such special emphasis on the earth? We notice that the inconceivably vast heavens with its innumerable stars is regarded as a unit and then mentioned together with the earth. This is clear from the expression *heaven and earth.*

The importance of the earth is not found in its size, but because it was made in such a way that it could sustain living beings, and ultimately also man. Thus the earth is different from the other planets in the galaxy. In all probability, it is unique within the entire universe. (Attention will be paid to this in the next chapter.) The following facts are crucial in this regard: the distance between the earth and the sun; the earth's mass; the revolution of the earth on its own axis; its elliptic orbit around the sun; the atmosphere around the earth; the separation of the land masses from the oceans; the raw materials which produce the food for plant and animal life; the ecological system that provides the necessary balance for the survival of living beings et cetera. The earth's particular significance is found in the fact that God, the Creator, is participating in man's history on this planet. In particular, earth's importance is stressed by Christ's coming to this earth. Here he became flesh and the Saviour of mankind, as well as of the entire creation (Rom 8:19–25).

15 David L Block declares in this connection: "Today we know that if we were to count all the stars of the Milky Way at a rate of one every second, the process would take two thousand five hundred years! ... A realistic estimate of the number of galaxies in our visible universe is conservatively 10^9 (1000 million)." *Our Universe: Accident or Design?* 1992, pp 6, 23.

2.2 The importance of man

We have already pointed out the infinitesimal minuteness of the earth in comparison to the other celestial bodies. Likewise when one beholds man, one is struck by how small he is in comparison with the earth on which he exists. However, according to the creation narrative, man is the climax and *the* special being who originated during the creation.

The importance of man does not reside in his size. It is found in the fact that he has been created in the image and likeness of God and that he stands in a direct and personal relationship to his Creator. Man is that which all the mighty celestial bodies together can never be. They are all merely matter while man is not only a living being, but he is also of *spirit* just as God who is *Spirit*.

David stressed this truth when he penned the following words: "What is man, that You are mindful of him, And the son of man, that You visit him? For you have made him a little lower than the angels, And You have crowned him with glory and honour. You have made him to have dominion over the works of Your hands; You have put all *things* under his feet," (Ps 8:4–8).

Man's importance is confirmed by Christ's incarnation here upon earth in order to become like man. In that way He saved man so that he could be like Christ for ever.

3 CREATION AND THE FALL

Everything made by God was perfect and good. We are, however, witnesses of the disruptive, destructive, chaos-creating and life-destroying forces at work in this creation. These can even be detected in the life of man who was created in the image and the likeness of God.

Earlier chapters (those dealing with hamartology: the doctrine about sin) have already pointed out that this was as a result of the Fall. The question which now looms large is: When and in what way did the Fall influence creation?

3.1 Creation and the Fall of man

Some maintain that the disruption, the suffering and death encountered in creation is as a result of the Fall of man. The Scriptural basis to uphold this statement are the words of Genesis 3:17 "cursed is the ground for thy sake".[16]

However, if certain facts are kept in mind, then one cannot merely maintain that the entire creation came under judgement on account of the Fall of man.

The curse pronounced in Genesis 3:17 was intended to punish man. The curse implies that circumstances on earth would be such that the existence of man would be accompanied by trouble and difficulty. Where everything in the Garden of Eden was all harmony, peace, safety and pleasantness, man now found himself in a situation where his existence was threatened in diverse ways.

16 Matthew Henry's Commentary puts it thus: "The ground, or earth, is here put for the whole visible creation, which by the sin of man, is made subject to vanity ..." p 32. Cf also H Ridderbos, *Aan de Romeinen,* 1957, p 186.

The mere fact that God provided the Garden of Eden for man to dwell in gives us every reason to accept that outside of that Garden, that is on the rest of the earth, things were not so ideal. The influence of a judgement must already have been operative there. Linked to this notion we should pay attention to the task set by God for man in the Garden. Besides tilling the soil, man had to guard it (Gen 2:15).[17] This presupposes a threat from outside the Garden. The existence of such a threat suggests that the forces of chaos and death were already present on earth.

The fact that Satan came to man in the form of a snake in the Garden to tempt him to sin indicates that the fall of the angels had already taken place. It presupposes that before the Fall of man, Satan already exerted his evil influence on creation.

Death was the punishment which followed the transgression of the so-called *test command,* that is, the command that man should not eat of the tree of the knowledge of good and evil (Gen 2:16–17). In the light of what one reads later in the Bible about the Fall, it primarily implies that man was estranged from God, the source of life. Thus man died spiritually. However, it also implies – and thus it has to be understood according to the test command – that man would die physically. When man transgressed this command, he did not immediately drop dead. What happened is that man came under the power of death from that moment onwards. The power of death comprises everything which threatens man's existence. Included in this are the natural phenomena which cause misfortune, disaster, suffering and death. It also means that death is operative in man's relationship with his fellow man. By this we mean the fact that people hate one another, subject one another to suffering, make one another ill and even murder one another. In particular, this means that man who was created from the dust of the earth would return to dust (Gen 3:19).

Sin did not only imply death and judgement for the first human couple, but also for their entire progeny(Rom 5:12). That which the narrative conveys to us about man's fall is not what happened to creation, but how it affected man. As regards creation, we notice that Paul mentions that it was subjected to futility in its entirety.[18] As a result of this, creation cannot reflect the abiding and perfect glory of God. It is subjected to the power of futility (death) by means of which it is kept captive. It is only at the final end and the restitution of all things that creation will be delivered from this bondage. Only then will it, in the form of the new heavens and the new earth, come into " the glorious freedom of the children of God." (Rom 8:21.)

As regards man's fall and its impact on creation, we only read about the thorns and thistles springing up on earth which would cause man trouble (Gen 3:17–19). Moreover, it is unthinkable that the entire creation, including

17 The Hebrew word *shamar* means: "to keep, watch, guard, protect." B Davidson, *The Analytical Hebrew and Chaldee Lexicon.*

18 The Greek word for *vanity (mataiotes)* has also the meaning of useless, futile, empty, futility. As an adjective *(mataios)* we also have the meaning of something which leads to nothing, which is foolish (Rom 1:21; Eph 4:17; Tit 5:9) and futile (1 Cor 15:17; Jas 1:26). All these speak of the influence of a power which creates chaos. Cf Louw and Nida, *Greek-English Lexicon Vol 1,* 65.37, 65.38. Compare also Bauernfeind, *Theol Dict NT*, 4, pp 522, 523.

all the heavenly bodies, could be affected detrimentally by man. The power which subjected the entire creation to futility and transience must be much greater and more powerful than man on earth. It had to be a power which could exercise more influence and control over the entire creation. This implies a power which was placed very high in God's hierarchy of creatures.

Considering everything mentioned above, one can only arrive at the following conclusion: The fall and the judgement by means of which creation was subjected to futility and transience must have already taken place before man appeared on the scene and fell to sin.

This view about the earth is confirmed by extra-biblical sources. According to palaeontological findings, disruption, suffering and death amongst the animals occurred long before the first man appeared on the earth.[19]

3.2 The creation and the fall of angels

From our discussion so far, it appears that the only logical conclusion to arrive at is that the entire creation was affected by the fall of certain angels. These angels apparently were in such a relationship to creation that their fall and consequent judgement affected it detrimentally. Since the Bible does not speak about this directly, this view is merely a possibility.

In both the creation narrative and the narrative of the Fall of man, no mention of the fall of the angels is made. Genesis only states the existence and the evil influence of Satan as a fact, without explaining its origin. Only much later on in the Bible do we read about the fall of some of the angels (Jude 6). We do not know when this actually occurred. Apparently it occurred during the period between the beginning of creation and the appearance of the first man on earth.

Let us accept the link between the fall of the Devil and his angels on the one hand and the subjection of the visible creation to futility and transience on the other. In addition to this we are informed that creation is not in this state on its own account, but because of an external force (Rom 8:20). The question now is: Who subjected this creation to futility and transience? Was this inflicted by evil angels, or did this happen when God subjected these angels to judgement?

We know that Satan has the power to create chaos and death on earth. However, this only happens within the limits set by God. In this regard we can consider the experiences of Job. However, we also know that God judges unrighteousness and that the curse can have an impact on nature in this way (Gen 3:17; 6:1–17). In terms of this, the Lord assures us in Revelation 22:3 that there will be no curse on the new earth.

In conclusion we may state that God allows evil to operate within certain limits in the visible creation. At the same time we also know that the Lord rules over all and that everything will eventually redound to his glory. This promise of redemption from futility and participation in the glorification is

19 With reference to this, E Sauer writes: "Death in the primeval world is not a geological hypothesis but a palaeontological fact. For anyone who knows anything of the facts it is completely incontestable that fossils that are definitely pre-Adamic exhibit traces of death and decay, indeed often of horrible destruction of life." Op cit, p 69.

also valid for the visible creation (Rom 8:21–23; 2 Pet 3:13; Rev 21 and 22).

In this promise one once again detects the prophetic element in the Genesis narrative. That which happened in the distant past is inseparably connected to God's work of re-creation through Jesus Christ. It ends in his eternal kingdom where everything will be subject to Him.

4 CREATION AND GOD'S PROVIDENCE

As already mentioned, creation came about by the Word of God. Without the Word there could have been no creation. Without this same Word it cannot continue to exist. Creation can never be seen as an autonomous entity apart from God.[20] It exists exclusively for God and for as long as He wills.

God did not create everything and then abandon it. Creation is not comparable to a clock made by a watchmaker, who wound it up and allowed it to run down. Such a deistic view is totally foreign to the Bible.

In Holy Scripture we not only see how God maintains creation, but how He also determines the course of its history. Together with this He grants limited opportunity to mentally advanced and morally accountable beings (angels and men) to exercise the competence of their own will.

God's involvement with his handiwork is known as his providence. We shall focus on important points in this regard and in particular, on those which have to do with man.

4.1 God sets limits to the influence for chaos-generating forces over creation

Although God allows Satan and his angels to influence creation negatively, their power still remains limited within certain boundaries. God ensures that creation does not end in chaos and that all life is not destroyed. He maintains all things (Ps 148:5–6). However, serious distortions occur in creation and we have to contend with natural disasters. These, however, never assume such proportions that human existence becomes impossible. Although creation has been subjected to futility, earth remains the sphere where God fulfils his purpose with man and this world.

The chief assault of the Evil One is on the Church of the Lord. However, He ensures that the forces (gates) of death never overcome it (Matt 16:18). The believer is fiercely attacked by the Devil, but God determines that he is never tempted beyond his ability so that he may withstand every onslaught (1 Cor 10:13). In addition the Lord has enabled man to combat many of the chaotic and death-bringing forces. As mentioned in Genesis 3:16–19, man's existence on earth is accompanied by trouble and even suffering. Yet the Lord enables man, albeit with effort, to secure his earthly existence and even experience prosperity in certain areas. This is because God did not allow man to lose his humanity when he fell. In concert with this, he maintained the ability to rule the earth and live creatively on it. Thus, by means of science, man is able to combat the things which threaten him on earth, such as natural disasters, climatological changes, sicknesses, pests and

20 It was the notion of ancient Greek philosophers that the visible creation existed always, and that it is the area where plants, animals, men and gods find themselves.

so on. Even though he is unable to eliminate many of these negative forces, he can at least minimise their detrimental effect. We see in all this how God maintains his covenant with creation and, in particular, with man (Gen 9:9–17). In this way we are assured that everything, yes, even mankind, will continue to exist until Christ's return.

4.2 The divine order in creation

God has created an innumerable number of material objects, together with an immensely great variety of organisms and living beings. In the universe in which they exist in concert, they together form a marvelous unity. Together they exist in a single great and incomprehensibly wonderful ecological system, with everything mutually dependent. Just as everything is in relationship to one another in the macrocosm (heavenly bodies and this earth) and thus ensures their survival, so also in the microcosm. The composition and functioning of the atom and the different molecules speaks of a plan and order which fills one with holy admiration. It is little wonder that one of the ancient prophets said of God and his creation: "He hath made the earth by his power, He hath established the world by his wisdom, and hath stretched out the heavens by his discretion." (Jer 10:12.)

We also see this predesignation of creation in the way in which the Lord determined a time and place for everything under heaven (Eccles 3:1). He knows everything and is involved with everything, yes, even those things that are most insignificant in the eyes of man (Heb 4:13; Matt 10:28, 30).

The inconceivably mighty power and dynamics present in the universe works according to divine will and plan. The Creator can also alter its working according to his will. We shall see this especially when we study the last things (Matt 24:29).

Divine order is particularly striking, seeing that all these forces in creation function according to scientific laws. We do not find just a vast amount of energy and dynamic in the immeasurable universe without any orderly structure by which everything might be plunged into chaos. No, everything functions according to natural laws in a single system.

Looking at the earth, we perceive how one dynamic system links up with another. We note how the sea is always in motion, and it affects both life in the sea and on land. We also perceive that everything operates within boundaries (Prov 8:29).

This operation according to the laws of science and the constancy due to the arrangement in creation, not only makes life on earth possible; it also ensures its survival. We think of the winds, the rain, the seasons, the alternation of day and night, the interaction between plant and animal life, the role of the sun and the moon on the earth and so on. Everything bears witness to a God of order who created and sustained it all (Isa 45:5–8).

We recognise the wonder and the necessity of order in creation (laws of nature), in the processes of procreation, in the genetic processes, the adaptability of organisms to changing circumstances, the way in which living beings are programmed with regard to danger, obtaining food, the building of shelters et cetera. This is also seen in the wonderful food chain, as well as in the balance in nature. Within the ecological system everything on earth plays a significant role.

Every academic discipline presents us with something of this wonderful order in creation. When we view all this in the light of Holy Scripture, then we see the way God cares for everything (Ps 104:27, 28; 145:15, 16; Matt 6:26). Without this order, creation could not serve God's plan, there could be no life on earth, and man's survival would have been impossible.

4.3 The divine order which is specifically applicable to man

God did not create man, endow him with certain abilities, place him on earth and then relinquish him to his own fate. It is not a matter where man is merely handed over to the forces and laws of nature, as is the case with the other living creatures. Man is intended to live in close relationship with his Creator. This is a relationship in which he would continuously experience the loving providence of the Lord. With a view to this intimate walk with God, the Lord placed the first human couple in a specially prepared place on earth, namely in the Garden of Eden. There man would converse with his God, experience his wonderful presence and rule as his co-worker on earth. This ideal circumstance in which man was placed tells of God's special provision for him.[21]

Before the Fall, God ordained certain regulations for man. In the first place He ordained that man should rule over the earth (Gen 1:26, 28; Ps 8:4–7). Therein lies the provision that man should practise science and technology. In this way man would come to greater self-realisation, whilst improving the world for himself, making it more pleasant and more beautiful. While man was in the Garden of Eden, the Lord also ordained that man should till the soil and guard it (Gen 2:15). Man's work and creative endeavours would contribute to his own spiritual and physical well-being.

Furthermore, the Lord also decreed that man should be fruitful and fill the earth (Gen 1:28). By populating the earth with good people, a richer, more meaningful and happier life would be ensured for all.[22]

As has already been mentioned, God did not withdraw his providence from man after the Fall. He gave man certain regulations which would assist him as a fallen sinner to live in a right relationship with God, his fellow man and the world in which he found himself. We see this particularly in the law of God in its outward form which He gave to his people, Israel.

The ordinances of this law were not only intended to teach man obedience to God. It also intended to eliminate those things harmful to man, and to further that which is in his best interest. When we pay attention to the law of Moses, then we see that it applies to the right relationship between man and God and also between man and man (the great commandment of love).

21 In ancient holy writings, traditions, legends and myths of different nations across the whole world, we find references which point to a garden in which the first human couple lived. It is a place where there is no suffering and strife, where man and beast lived together in harmony, where men had fellowship with gods, where the tree of life was, where men committed sin and similar things. All these have some resemblance to the Biblical narration of paradise. Cf G Ch Aalders, op cit, pp 341–411.

22 E Sauer declares that God put man on earth with the definite purpose to deliver it from the power of The Evil One. Op cit pp 92–99.

The correct notion, however, is that Christ and not man, would ultimately deliver creation from all vanity and corruption.

It also deals with the protection of life and the individual's basic rights. It acknowledges and protects institutions such as marriage, the family, spiritual and secular authority, and so on. It contains regulations about health, the right food to eat, the correct way of utilising the earth, the treatment of animals and so on. It also contains regulations on combatting crime and perverse acts. Similarly there are other stipulations which show that God seeks only man's welfare on earth.

In the New Testament the gospel is presented very clearly. In this we find the law of Christ, and how true life for man is presented and made possible.

To the extent that a country's laws regarding society, politics, economics, cultural and other matters concur with the biblical regulations, to that extent does it reflect the providence of God.

In all these ordinations it is clearly stated that those who keep them will be prosperous, whilst those who disregard them will be punished.

4.4 God's providence in his work of re-creation

After the Fall of man, God immediately made provision for man's redemption. We see this in the mother-promise in Genesis 3:15 (cf also Rom 16:20 and Heb 2:14–15). In this promise, judgement is pronounced upon the snake (the Devil) and mention is made of the One who would be born of a woman in order to save mankind from futility and judgement.

This promise of the coming Messiah is found throughout the Old Testament. It is also linked to the promised new covenant (Jer 31:31–33; Gal 3:16). This was the hope which shone like a bright light in the hearts of believers. This promise was consequently fulfilled when Christ came to earth to bring about reconciliation between God and man. The gift of God's Son speaks of an unutterably great and gracious care for man. Christ is God's actual provision for a fallen mankind.

This provision does not only exist in the eventual complete redemption of believers. It is a reality and a power which is presently at work. One sees this especially in the re-creation already taking place in the lives of people. In accordance with this, man becomes a new creature (2 Cor 5:17).

The Lord never abandoned man after the Fall. As with the first human couple, He still visits the lost. He does this through the working of the Holy Spirit who makes various ministries possible through people. Thus there have been, from the beginning, those who preached the word of salvation and hope to others. We think of the preachers and shepherds of the Old Testament in the persons of priests, prophets and kings. We also read of such ministries in the New Testament (Eph 4:11–16).

God's history with the believers is recorded throughout Holy Scripture. Here we read how God, just like a faithful shepherd, has cared for his own flock and provided them with good things (Rom 8:28).

The influence of the work of re-creation accomplished by God in the lives of believers is validated in this creation. The fellowship of believers is like light which dispels the darkness and like salt that preserves from decay (Matt 5:13–16). True believers set a good example to be followed by others. By their stand for that which is right and good, they serve as a conscience to the society in which they find themselves. For the sake of those who belong to the Lord, blessing and expedience also come to others (2 Kgs 6:15–23; Acts 27:23–24; Gen 18:23–32).

The Lord's provision in this creation is also visible in the miracles which He performed. We see this in the way he delivered Israel from Egypt; the way in which He led them through the desert and the way in which He drove the enemies from the promised land. We also see this in the gifts of grace bestowed by his Spirit (1 Cor 12:1–11) whereby He intervenes in this creation in order to help man in his need. Miracles combat the influence of sin in the life of man as well as in nature. They confirm that which is good and destroy evil and deceit.

God's work of grace in the history of man is a confirmation of his providence in this creation. His provision will reach its climax and fullness when the present creation is replaced by a new heaven and earth (2 Pet 3:13).

CHAPTER 3 The biblical doctrine of creation and sciences – the purpose and significance of creation

Besides that which the Bible teaches about the origin of all things, we should also take cognisance of what is said by the sciences, such as astronomy, geology, genetics, palaeontology and others. These sciences make interesting pronouncements about the origin of the universe and, in particular, about the earth's origin. Furthermore, much is said in connection with the origin of living creatures.

When one makes a comparison between biblical teaching and that which is presented by science, then it may seem as if there are contradictory statements regarding certain things. As was already mentioned, we do not believe that there should be conflicting ideas. That which God teaches through his Word and that which He created (and which is studied by the sciences) is part of one and the same truth.

If the biblical doctrine of creation and scientific discoveries seem to clash, there must be a reason for it. Either the Bible has not been correctly interpreted or incorrect conclusions have been drawn from scientific discoveries, or both.

To begin with we must state unequivocally that we accept the Bible as the basis for the existence and the origin, the meaning and purpose of the entire creation. Scientific knowledge can, however, be used in a supplementary way. The Bible does not give us a complete nor a scientific overview of the origin of all things. It only provides us with a few major truths which are necessary as a point of departure for the history of mankind and God's message to men. On the other hand, science can give us details which are not dealt with in the Bible.

We have to state very clearly that the biblical view should not be adapted to suit any scientific pronouncements. The validity and the authority of any biblical doctrine (theology) is only determined in as much as it interprets Scripture correctly. However, science may assist in checking any incorrect conclusion drawn from Scripture.[1]

Science may also be helpful in that it provides us with insights, terminology and expressions whereby the biblical doctrine of creation may be presented in a way relevant for modern man.

1 In this regard C J Dippel declares that natural scientific experience is able to call wrong theology to order. True theology, however, has to do all in its own strength. *Kerk en Theologie,* Julie 1965, p 126.

1 COSMOLOGY: BIBLICAL AND SCIENTIFIC EXPLANATIONS

In the first place, we focus on what is presented as the biblical view of the origin of the universe (cosmos). Thereafter we also focus on what has been said from a scientific point of view.

1.1 A total view of the universe

There was a time when theologians advocated a *geocentric* cosmology (earth-centered doctrine of the universe). This view, however, does not have its origin in the Bible, but comes from the ancient Greek and Roman philosophers. The Aristotelian idea of the universe, as further developed by Ptolemaeus (90–168 AD) explains that the earth is the centre and fixed point of the universe. The sun, moon, stars and other planets revolve around the earth from east to west, each in its own orbit. This is the geocentric view. It was regarded as the biblical view, chiefly by the theologians of the Middle Ages.

Nicolaus Copernicus (1473–1543) was one of the first who presented a more scientific cosmology. He argued that the earth revolves around the sun, thus presenting a *heliocentric* view. He laid the basis for later astronomy. He was followed by people like Bruno, Kepler and Galileo. The latter two were persecuted by the Church since this view was regarded as a heretical deviation of the biblical doctrine.

Ultimately the scientific view triumphed. Together with this, however, rationalism (especially under the leadership of Descartes) also gained tremendous acceptance. People began to turn their backs on the Church and rejected the so-called biblical cosmology as prescientific. Only when Isaac Newton (1642–1727) appeared on the scene did matters change in some way. Newton was known for his excellent work,such as the discovery of the law of gravity and its implications for the universe. He unequivocally professed that the cosmos cannot be explained without the idea of God who created everything.

Gradually people realised that the geocentric view of Aristotle could not be deduced from the Bible. It was a case of theologians reading in the Bible that which it did not actually propound. As a result of this wrong notion they declared, without more ado, that the biblical view of the cosmos could never be reconciled to any scientific cosmology.[2]

The truth of the matter is, however, that the Bible does not give us a world view or a cosmology as some would like to believe. Where, for example the Bible says that the Lord "established" the earth forever, it is not stated in connection with the universe (Ps 78:69; Prov 3:19). These instances refer rather to the security experienced when we stand on the earth. This is subsequently linked in a poetic way to God's faithfulness, steadfastness and wisdom.

According to Joshua 10:12, Israel's leader prayed for the sun to stand still in order to lengthen the day. This is based on the idea that the sun rises over the earth and then sets again. In this case we have an expression still used today which should be understood in that context. This does not necessarily imply a *geocentric* cosmology.

2 Cf E Brunner, *Dogmatik*, 1, p 51.

On the other hand, we also find biblical pronouncements that can be linked to the present view of the universe. Thus we read in Proverbs 8:27 that the earth must be spherical. And again, we deduce from Luke 17:34–36 that day and night can occur on earth at the same time – an indication that the earth is suspended in space and that it revolves around its own axis.

However, one thing is certain, namely that the Bible does not give us a description of the universe and how all the heavenly bodies relate and function with regard to one another.

1.2 The significance of a scientific theory

The Bible is not in conflict with things that have been proven scientifically beyond any doubt. However, a problem arises when certain scientists put forward all kinds of theories and philosophical speculations which are neither biblical nor scientific.

It is acceptable in the natural sciences to postulate a theory with a view to research. One then works with such a theory until proven true or false. After this it is accepted as a scientific fact or it is rejected. Such a theory should, however, always remain within the boundaries of the sphere of a scientific approach. It can, for instance, not make pronouncements about metaphysical speculations which replace God and his work of creation in an aprioristic sense. Such speculations can in any case never be tested by the natural sciences.

Genesis 1:1 says: "In the beginning God created the heaven and the earth." Man has to either accept or reject this depending on his religious stance. This statement cannot be tested by the natural sciences. In any case, this falls outside of the purview of any scientific research. The natural sciences study that which already exists. They study its nature and function as well as its possible history. (We will return to this when evolutionism is discussed.)

1.3 The beginning and the development of everything that exists

According to the Bible everything we observe had a beginning, hereafter it developed to what it is today. We read about this beginning in Genesis 1:1. In the rest of the chapter we read about the process of development which things underwent, especially that of the earth until the appearance of man. Science confirms this truth.

The universe as observed by us did not exist since all eternity. It had a beginning and a history of formation while changes are still taking place. As was already mentioned, astronomers explain that everything which the universe consists of was probably contained in one object. The origin of this object or mass falls outside the parameters of natural science and transcends human intellect. It is further postulated that this original object overheated and exploded. This is usually referred to as the *Big Bang* which took place some 15 billion years ago. With this explosion particles were flung into immeasurable space in all directions. (Where space comes from once again falls outside both man's conception and science.) As these particles travel through space, they move yet further and further away from each other. Thus we have to do with an ever expanding universe. The poetic words of Isaiah 42:5 can be linked to this: "Thus says God the Lord, Who created the heavens, and stretched them out, ..." (cf also Isa 40:22).

With this explosion there was certain matter (plasma) which glows and emits light, while other matter was dark, possibly on account of cooling. By means of a series of processes a planet such as our earth originated. Primitive life appeared first on this planet, then later higher forms of life and then eventually man. What scientists learn from space research supports this theory.

What is important in this case for the study of dogmatics, is that the heavens (together with the earth) were not always there. It had a beginning and thereafter underwent a process of development. This process of development as well as that which took place on earth serves as a confirmation of what is read in Genesis 1:2–26.

1.4 The laws of nature and God's provision

According to cosmology and astrophysics, not only matter, but also the laws of nature as encountered here on earth are the same as those occurring in the rest of the universe. Here we refer to phenomena such as the law of gravity, chemical reactions, chemical compositions, atomic compositions, atomic reactions, the way in which light is brought forth, how the rays of light of different elements differ from one another, the speed of light and all kinds of factors that influence light rays, as well as other laws and reactions known to physics. The law of gravity, electrical power and nuclear power hold the universe together in a unit and determine the behaviour of the heavenly bodies as well as their movements in space. By heavenly bodies we mean galaxies, quasars (very bright masses of gas occurring in the spaces between galaxies), *black holes* (objects with tremendous gravity which emit no light), stars, planets and moons.[3]

In accordance with their available knowledge, cosmologists are trying to form an idea of the possible process which took place from the time of the Big Bang until the appearance of life on earth. Some of the steps that are postulated in this process will now be enumerated.

At the time of the Big Bang, masses of matter were flung into space. This plasma together with its gas content must have gradually been transformed into heavy elements and long-living galaxies and stars. On account of thermo-nuclear processes in the heart of some stars, elements and molecules necessary for life were formed. Here we think of carbon, oxygen, nitrogen, phosphor and so on. Some of these stars would then have exploded and changed into supernovae. As a result of this explosion the elements necessary for life landed in space. In turn these were absorbed by the gaseous cloud from which our galaxy originated. In this way some of these elements became linked to the earth where they provided the building blocks for the development of life.

In order to make these steps possible, a series of incidental events had to take place. Thereby we mean things which cannot be explained in terms of the inherent forces and laws of the cosmos itself. Above all, the different

3 Most of these data are derived from *Cosmic Coincidences* by John Gribben and Martin Rees.

incidents had to be controlled by an all-embracing or master incident. This had to ensure that everything did not lapse into chaos, but that everything linked to one another in orderly systems in order to eventually form one great unit.

We can accept that further cosmological research and insights can lead to a scientific explanation of some events now seen as incidental. Something like this again emphasises the wonder of the laws of nature which were incorporated into the primeval plasma from the outset. This in turn indicates that all things must have been programmed in this way beforehand. This also indicates a purpose for all the cosmic events to, among others, bring about life on a planet such as the earth.

The Bible-believing scientist who accepts the biblical doctrine of creation should not have many problems with this kind of scenario. It can be regarded as mere detail in connection with the creation which are not mentioned in the Bible, but which yet, if proven beyond doubt, can be aligned with the major truths about creation.

Once again we must point out that the Bible does not give us details about the creation of heaven and earth in the way that the astronomers surmise it might have happened. For this reason these particulars cannot be included as part of the *biblical* doctrine of creation. In this regard it is also significant that demonstrable cosmological data indicates that superhuman wisdom, power, planning, ordering, et cetera has to be linked to creation.[4] For those Bible-believing people, this can be seen as a pointer to the Creator. It discloses to them that there is a Creator and Sustainer of all things, and that he has a definite plan and purpose for everything. Moreover something like chance does not exist for the believer. All these things which have been mentioned happened according to God's purpose. He alone determines the course and destination of all creation.

2 THE ORIGIN AND DEVELOPMENT OF LIVING CREATURES ON EARTH

The origin and the development of living creatures is an area where biblical faith and science are related in a special way.

In our discussion of this topic, we have to state at the very outset that biblical faith should not necessarily have a problem with the notion of a developmental process with regard to the creation of living creatures. Certain fundamentalists, however, reject any development of this kind.[5] They believe that everything that we observe was originally created by God immediately and precisely the way it is now.

4 In this connection we mention the following statement: "Dr A Cressy Morrison pointed to many facts commonly known to science which, in his words, prove that our universe was designed and erected by a great engineering Intelligence." Quoted by Paul A Zimmerman, Darwin, Evolution and Creation. *Reader's Digest*, Dec 1946.

5 This is the notion of G Ch Aalders. Op cit, p 153. Referring to the days of creation, L Berkhof declares: "But some of the Church Fathers, who intimated that these days were probably not to be regarded as ordinary days, expressed the opinion that the whole work of creation was finished in a moment of time, and that the days merely constituted a symbolic framework, which facilitated the description of the work of creation in an orderly fashion, so as to make it more intelligible for finite minds." *Systematic Theology*, p 278.

2.1 Gradual development as the way in which everything came about

However, we maintain that there is sufficient indication in the Bible (and in nature) that God also brought things about by means of a process.

The Bible says expressly that everything as we observe it did not reach its final stage in its first moment of existence. First and foremost we see this in the way that the earth was prepared until the point where living beings could exist on it. We also read how different living creatures came into being according to a particular sequence. A particular day (a period) of creation is specified for the origination of each group. The words: "let the earth (or waters) bring forth" plants, fish, animals, etc, is also an indication that things developed with the earth or the sea as matrix (Gen 1:11, 20, 24).
Moreover, in nature, God has determined that plants and trees grow to maturity from a seed or a shoot. Animals develop from a fertilised ovum to their full stature.

We also encounter the developmental nature of God's actions in his work of re-creation. The Lord uses the process of growth and maturation as seen in nature to describe the establishment of his kingdom on earth (Mark 4:26–32; Luke 13:18–21). The entire salvation history as it unfolds in the Bible tells of a process which stretches from the first promise (Gen 3:15) to the promises and covenants made with the patriarchs, throughout the history of the nation of Israel until the birth of Christ on earth. Thereafter, we have his ministry in this world, his passion, death, resurrection, ascension and the outpouring of the Holy Spirit.

Even the re-creation of man takes place as a process. Firstly man is reborn by the Spirit of God. This is something that happens in his heart or spirit. Sanctification follows, then death and the resurrection from the dead, together with the obtaining of the resurrection body and ultimately the glorification with Christ in the hereafter.

2.2 The notion of evolution

In connection with the origin and development of the different living species, the biological sciences make use of the word *evolution*. This is derived from the Latin *evolutio*. This again is connected to *evolvo* and *evolutum* which means: unfurl, unfold. Evolution expresses the idea of unfolding and *development* whereby that which was first concealed in a small and simple form gradually develops fully according to its inherent potential. In the biological sciences this refers in particular to the series of gradual progressive changes in an organism on account of both inherent forces as well as the influence of external factors. In this way a simple (low level) organism would develop into a much more complex and higher level of life. On account of the workings of inherent forces, adaptation to external circumstances, mutations, cross-breeding, the supplanting of the weaker by the stronger (so-called survival of the fittest), procreation of acquired characteristics, natural selection, and so on, organisms develop in different directions, yes even to the extent that new species or kinds originate.

The notion of evolution emanated from the observation of the many similarities among different species in the plant – as well as animal kingdom. Here we refer to morphological, embryological, reproductive, physiological, neurological, endocrinal and genetic characteristics. The fossils of prehistoric living creatures which have been researched by palaeontology contributed particularly to the theory of evolution. This research has uncovered indications of similarity between the different species. This kind of data points to a universal and similar life mechanism and to the unity which exists in both the plant – and animal kingdoms. The notion of evolution existed from very long ago.[6] However, it was really only from the time of Charles Darwin's publications that the theory of evolution began to gain wider acceptance in scientific circles.[7]

According to evolutionary theory man developed from the same primeval ancestor as the apes. The human branch as distinguished from that which brought forth the apes is known as pre-hominids. They developed into the present man *(homo sapiens)* over millions of years. During this evolutionary development different branches of the pre-hominids led to the origination of the five basic races, namely the Mongoloids (yellow peoples), Negroids (black peoples), Austroloides (natives of Australia), Capoids (the Hottentots, Bushmen, etc.) and Caucasoids (white peoples).[8]

2.3 The notion of evolution and the Bible

It has already been pointed out that we are of the opinion that the idea of the development of living creatures does not necessarily clash with biblical statements. A problem only arises when that which is said about evolution is no longer strictly limited to the area of scientific inquiry. On the one hand we read in the Bible that God made everything in the plant and animal kingdoms "according to its kind" (Gen 1:11, 21, 24). On the other hand we have to keep in mind that what we read about in the Bible describes the final product. Therefore the possibility of a process which took place during the course of a day of creation and which could have led to that final product, is not excluded. Several theologians accept that an evolutionary process was part of God's work of creation.[9]

2.3.1 A theistic doctrine of evolution and the Bible

An attempt to reconcile the theory of evolution and the Bible with regard to the origin of man, is found in the theistic teaching on evolution. Dr Hans

6 S du Toit refers to the following historical facts: "The Greek Anaximander (ca 550 bc) was the first author who occupied himself with the question about evolution. Empedocles (also a Greek, 495–435 bc), however, is considered to be the father of the evolution notion. He was of the opinion that nature, through several events, tries to produce more perfect beings, whilst eliminating the imperfect ones." *Bybel, Skepping, Evolusie*, 1968, p 278.

7 We specifically refer to his publication: "On the origin of species by means of natural selection," which was published in 1859.

8 The Ancestry of Man, Evolution, Life Nature Library, *Time-Life Int*, 1964, pp 153–175.

9 S du Toit quotes what Van Selms wrote about evolution: "Van Selms asserts that God had called more than once for the co-operation of his creation: let the earth bring forth ... !' This must not be understood as if the Bible teaches that living plants were produced by dead matter, because without God's fiat the earth would not be able to do it. Van Selms is of the opinion that this offers a point of departure for a Christian doctrine of evolution." Op cit, p 197. A de Bondt declares: "Genesis 1 is a clear indication for us, that God started with that which was lowest. After that He created something else, ascending to that which was higher. But at the same time He ascended creatively. For this reason we must not place creation and development against each other. They belong together. Those who believe in evolution without creation, offend God, the Creator. Those, however, who have not an eye on the system, on the development, on progress, offend the multilateral wisdom of the exalted God." *Het Dogma der Kerk*, under the editorship of Prof G C Berkhouwer and Ds G Toornvliet. p 231.

Rohrbach is a proponent of the latter.[10] He explains that Adam, just like the plants and animals, developed from a lower form of life into the human form, according to the evolutionary principle. He deduces this from the first narrative on creation in Genesis 1. After Adam's biological development into a human, God chose him above all the other living creatures on earth. He breathed his breath into Adam and thus he became a spiritual being. The latter is based on the second creation narrative in Genesis 2.

Together with Adam other creatures also developed into man-like apes. This would be confirmed by fossil finds of such pre-hominids. Biologically speaking these other creatures did develop into humans, but were not made spiritual beings by God. Rohrbach refers to these creatures as para-Adamites. He finds the existence of such a para-Adamite race essential in order to provide answers to certain vexing questions in the Genesis record. Thus Cain had to be protected against this race; it provided Cain andSeth with wives and Cain built a city with members of this race (Gen 4). God was to have obliterated this para-Adamite race with the Flood.

There are serious biblical objections with regard to the notion of para-Adamites who were to have existed together with the first real people. The Bible says absolutely nothing about these animal-like people who existed together with the real people and with whom they were to have intermarried. The mention in Genesis 6:1–4 of the *sons of God* who married the *daughters of men* also provides no reason to believe that real people married animal-like people. (This should rather be understood as the intermarriage of believers with unbelievers which led to unrighteousness).

If Cain and Seth married para-Adamites, then it implies that we as people are in fact a mixture of the true man created in the image and the likeness of God and animal-like people. We say this on the grounds of the fact that Noah is descended from the line of Seth, and that Noah in turn was the father of the entire human race after the Flood. Nowhere does the Bible mention such a mixture of man and beast. What is clearly taught is that all people originate from Adam and bear his image (1 Cor 15:45–49).

If there were such para-Adamites at the beginning of human history, they, along with Adam and Eve, also had to undergo a Fall. We read in Genesis 6:5–7 that those who perished in the Flood did so on account of their grave sins. The Bible mentions no other Fall among men besides that of Adam and Eve. All people come under the impact of the Fall on account of the fact that they are descended from Adam (and not on account of anyone else's transgression.) The following references are mentioned: Genesis 3:16–19; Romans 5:14–15; 1 Corinthians 15:22.

If theistic evolutionism wants to speak about the origin of man within biblical context, then the idea of the existence of animal-like people has to be abandoned. Another answer must be found to explain the origin of the other people mentioned in Genesis 4 besides Adam, Eve, Cain, Abel and Seth. The most obvious explanation is that Genesis 5:4 says that Adam and Eve had sons and daughters'. During the course of the centuries and by taking cognisance of the longevity of people at that stage, thousands of offspring

10 Cf A E Wilder-Smith, *Man's Origin, Man's Destiny*. 1974, pp 275–278.

could have been born of the sons and daughters of Adam. They were living at the same time as the very first generation. In addition one should keep in mind the prophetic-historical nature of Genesis. It has already been pointed out that this narrative is not concerned with the recording of ordinary history as we know it. Only that historical evidence is used and only those historical characters are referred to, who are important to the transmission of the prophetic-historical message.

2.3.2 A non-evolutionistic view of the origin of man

A very general theological view is that at the same time that God breathed his breath (spirit) into man, he also created man's physical body. This view does not regard the body of man as something which developed by means of an evolutionary process over a very long period of time. Moreover, it is pointed out that parts of skeletons of pre-hominid or apelike people which have been discovered in the form of fossils in different places on earth, in no way imply that they were the progenitors of man.

However, there are many similarities between man and beast. This, however, should be ascribed to the fact that man and beast live on the same earth, their bodies were created from the same raw materials, they breathe the same air and are nurtured by the same earth, et cetera. This does not mean, however, that one species developed into another or that animals later developed into people.

According to scientists these fossils come from creatures which existed millions of years ago. Subsequently the question is raised: why did no other pre-hominid develop from lower forms of life during all those millennia? Did the evolution which led to the origination of man come to a standstill or was the latter only a one-time freak?

In order to reject the idea that man developed from lower forms of life, many scientists are quoted who doubt the scientific validity of many of the claims and deductions of the evolutionists. Their main objection to the theory of the evolution of man, is based on their interpretation of the Bible. They point out that man was made by God and did not develop from animals. Moreover, a special act of creation, unlike that of the other living creatures, took place with man in that he was made in the image and the likeness of God.

We can agree with these statements in principle except for the idea that the human body was necessarily formed instantaneously from the dust of the earth. We have already mentioned that it is probable that God brought about things during creation by way of a process. Thus evolution could have formed part of the work of creation.

Regarding the creation of man, we read (Gen 2:7): "And the Lord God formed man of the dust of the ground, and breathed into his nostrils the breath of life; and man became a living being."

It seems from this verse that the entire human body was formed out of elements from the earth. Subsequently God breathed his breath (spirit) into man and made him a living soul. Depending on one's presuppositions, one can make more than one deduction from this verse. Firstly, it may be said that God formed the human body from the earth and that initially it was lifeless. On the other hand, it may be argued that God caused the human body to originate as a living organism from the earth in an evolutionary way. He then breathed his breath (spirit) into this living being who then became man. Since the Lord has not revealed this kind of detail to us, no final conclusion about this can be drawn dogmatically.

It is worth noting the Bible – here, like elsewhere – does not say so little that we do not know who and what man is. On the other hand, it also does not give too many details of a scientific nature which, in any case, would have been incomprehensible to the first readers. The Bible only provides us with the major truths which can serve as firm beacons for people of all times.

3 THE DISTINCTION BETWEEN MAN AND THE OTHER LIVING BEINGS

According to the creation narrative, as found in Genesis, man is no superhuman being who was transplanted onto this planet. He, like all other living beings, had his origin here on earth. The question is, however, how should we see man's relationship to other living beings? Is he just another kind of animal, or is he a totally unique being?

3.1 The essential difference between man and animal

The Bible mentions the similarity between man and beast. At the same time it also refers to the radical difference between the two. This difference does not so much reside in the area of physical appearance and skills. It exists in that man, unlike other living creatures, was created in the image and the likeness of God (Gen 1:26–27). As we have already seen, this implies, among others, that man, regarding his inner self, is spirit. This means that he is in a personal relationship to God and he exists as an *I*, precisely because the spirit of the I (God) is in him. Man's physical appearance also speaks of the resemblance to the image of God as revealed through Christ (2 Cor 4:4; Col 1:15). This image of God according to which man was created, is not something which came about by the process of evolution; it is eternal just as Christ himself is eternal. The purpose for which man was placed on the earth as well as his eventual destination, is totally different from that of other living creatures. (There are indeed also other differences between man and animal, but these are not important in this connection.)

3.2 Man as well as all the other living beings, testifies to God's handiwork

Contrary to the fundamental differences between man and animal, we should, on the other hand, also mention that man as well as all other living beings bear the imprint of the handiwork of God. It can possibly be said that inanimate things bear the least testimony to God's handiwork. The latter is more evident in plants and even more evident in animals. Finally it can be most clearly seen in man. Consequently man is presented as the climax and crown of God's work of creation on earth. In man, and in particular in a man who loves God and his neighbour, we not only see a clear indication of God's handiwork, but also of his image and likeness. (Indeed man, animal, plant life and all the other things in creation also show how sin marred God's handiwork. As regards this, the same tragic fate struck both man and the rest of creation.)

In summary, we can thus say that although man does show affinity to other living creatures on earth, there are radical differences. These do not only concern degrees of difference in certain abilities, but a difference in being. Outwardly or morphologically speaking, there are indeed many similarities between man and beast, especially between man and certain kinds of apes. Man as spirit, man in his distinctive relationship to God, makes him distinct from any other living creature on earth. Man is of such significance to God that He saves him from death and corruption by the redemptive death of his Son and makes him a participant in his eternal glory.

3.3 Man and the so-called pre-hominids

Having looked at the difference between man and other creatures, the question concerning the pre-hominids and man's possible kinship to them still remains.

We have already pointed out that the notion that man developed as an organism by means of a process, does not necessarily conflict with the Bible. In any case the Bible only says that the Lord God formed man from the dust of the earth (Gen 2:7). However, it does not describe how he did it. Could it be that the pre-hominids were part of such a process?

The Bible says nothing about the existence and the significance of the so-called pre-hominids. What it does refer to is man as known today. If God formed man by means of a developmental process, then it is only man as the final product with whom the Bible is concerned. We are the progeny of man created in the image and likeness of God. God destined a history until the end of this world for this man. For this reason the pre-hominids are irrelevant in this dogmatic study. However, by saying that we do not want to deny the existence of such beings (which later disappeared), all we are saying is that they are in no way associated with the conveying of the Bible's prophetic-historical message to us.

The matter of the hominids is nevertheless relevant when people make certain deductions which affect dogmatics on these grounds. For instance, it is suggested that man did not originate in any other way than the animals and man is simply on a higher rung of the ladder of evolution. The uniqueness of man, created in the image of God, has already been dealt with. (Anthropology, in Part One of *Words of Light and Life*.)

4 THE PURPOSE AND MEANING OF CREATION

We confess and believe that everything that exists was created by God and constitutes part of his self revelation. Therefore, we also declare that the universe, with everything which takes place in it, occurs within the ambit of God's will and plan. Among others this includes the freedom which he has given to self-conscious beings with an own will, who can choose to remain faithful to him or not.

By seeing creation as the revelation of God's will and plan, we also arrive at its purpose and meaning.

We deal with this topic from the following three perspectives, namely with a view to Christ, man, and God the Father.

4.1 The purpose and significance of creation with a view to Christ, the Word

We have already referred to God's creation of everything through Jesus Christ, the eternal Word (John 1:1–10; Eph 3:9; Heb 1:2).

4.1.1 The Word as the meaning and purpose of creation

The fact that God created everything through his Word means that he thus willed it so. This alone is sufficient reason *(ratio sufficiens)* for the existence, meaning and purpose of everything which was created.[11] The fact that God merely spoke and everything was called into existence testifies to his omnipotence and supreme providence. In other words, creation is an actualisation, a visible becoming of that which God spoke. Therefore, it also speaks of God's self-communication – He reveals himself through that which he brought forth. All this stresses the connection between Christ and the creation. It confirms that everything is from him, for him and unto him (Col 1:15–17; Heb 2:10). For this reason we can say the creation was created for Christ.

The creation made it possible that a bride and a body could be prepared for Christ. Here we refer specifically to the creation of man and the history of mankind on earth.

4.1.2 Creation bears the imprint of the Word

The Word through whom everything was created speaks of the revelation of the God of order, namely his will, his plan and manner of working. For this reason creation has to also bear the stamp of the Word in the form of order. A chaotic condition would simply not testify to God's handiwork.

We see this order in creation, for instance, in the laws of nature. Although the laws of nature have been provided by God, we may not equate them without ado with God's law or regard them as direct divine operation. The world (heavenlies and earth) always remains that which is not divine.

Furthermore, we find that nature does not always function with such inevitable rigidity as to enable man to make precise predictions. Take long term weather predictions for example. There are many divergent laws of nature, variables and possible events that impact upon one another and influence the existing order. For this reason, we may also speak of a chaotic and unpredictable element in nature. This contingence or variability is, however, always limited within a fixed system which ensures that everything does not lapse into total chaos.[12] We see something of the freedom and the contingence which God gave his creation, which at the same time is also the evidence of his gracious sustenance of everything.

11 Cf Emil Brunner, *Dogmatik*, 2, p 23.

12 With regard to this Colin E Gunton writes: "There is chaos but stability, that is to say, contingence but reliability. Chaos is ubiquitous; it is stable; it is structured. It is in fact, part of the dynamics of the universe. Those studying chaotic dynamics discover that the disorderly behaviour of simple systems acted as a creative process. It generated complexity: rich organised patterns, sometimes stable and sometimes unstable, sometimes finite and sometimes infinite, but always with the fascination of living things. The universe is not only contingent-free in its own way; but the contingency operates so as to be creative. Gleick's book *(Chaos: Making a New Science)* opens with what has become perhaps the pictorial symbol of the notion of chaos, the Butterfly Effect – the notion that a butterfly stirring the air today in Peking can transform storm systems in New York next month.' The fact that this concrete example illustrates even more remarkable and mysterious phenomena – for example, that minute subatomic changes in gases on the other side of the universe can affect the way things happen on this side ..." *The Promise of Trinitarian Theology*, 1991, p 155.

4.1.3 Creation leads to the re-creation, and both originate from the same Word

The Son as Creator (the One through whom God created) is also the Re-creator. Through the Son (Word of God) there is thus a continuity between the creation of the heavens and the earth and its re-creation in view of the ultimate consummation of all things (John 1:1–3; Heb 11:3; 2 Pet 3: 10–13). As creation came about by the Word, so the re-creation also comes about by the Word. The creation also provides the opportunity for Christ, the Word, to prove his love and obedience to his Father, through his work of re-creation. This begins with his redemptive work, and finds its climax and completion at the final end when there will be a new heaven and a new earth, and when the redeemed are with the Lord. This was all made possible when He took away sin by his humiliation, passion and death. In this way the grace and love of God could again find expression in the glorious re-creation.

4.1.4 Christ is glorified by the creation and re-creation

By means of the creation and re-creation the situation arose where Christ would not only be humiliated, but also exalted to the very highest. Thus he received a name above every other name (Phil 1:6–11). By means of the creation the situation also arose whereby Christ would engage in battle with Satan and his evil forces. He conquered these so that He might be revealed as the mighty Victor (Col 2:15). Every enemy would be subjected under his feet (1 Cor 15:25; Eph 1:22) and every knee would bow and every tongue confess that he is the Lord (Phil 2:10–11). As Conqueror He would snatch that out of the hands of the Evil One which he had usurped, and again establish the kingdom of God. In this way he would also be acknowledged as the King of kings (Rev 17:14; 19:16). By means of the creation and that which took place thereafter, all power in heaven and earth was given him (Matt 28:18). By means of the creation the opportunity arose where Christ would act as Judge (Acts 17:30; Rom 14:10; 2 Cor 5:10). He will also pronounce judgement over everyone whereby the eternal fate of the just as well as the unjust will be determined.

Every time the Lord acts in his creation, it results in his exaltation and glorification. For this reason we may say that the purpose and meaning of creation exists therein that Christ is exalted. Seen thus, creation speaks of God's love toward his Son, Jesus Christ.

4.2 The meaning and purpose of creation with a view to man

Certain truths become quite evident when creation is viewed in the light of man.

4.2.1 By means of the creation God brought forth beings who would bear his image

By means of the creation God has declared that he wants to possess us as people. He desires those who will bear his image and his likeness. Yes, because God created I am there. I am there not merely as another organism, but as an *I* (a subject) who exists together with and in relation to the *I* (God).

Precisely because God created me as a person – even though this happened through the lineage of the human race of which I am a part – I am a being who is too wonderful for even myself to comprehend (Ps139:13–14). By means of the creation God did not only bring forth people like us, but He also makes these beings participants in and heirs of that which is divine (Rom 8:17). That which belongs to his own divine nature He also shares with us out of his love and grace. The more He reveals of his Spirit in us, the more we can live to his honour and glorify Him.

By creating the world in which man has been placed, God gave him the opportunity to rule through the abilities planted in him (Gen 1:26, 28; 2:19). Ruling implies that man as subject can know the world around him as an object, investigate it through scientific endeavour and utilise it to his benefit – as well as to the benefit of nature as a whole (Gen 1:29; 2:15). God has made it possible for man to develop the potential placed within him through the practice of science, technology and culture. In this way man could develop his full humanity. He could act as co-worker, yes even co-creator together with God in order to maintain that which is in the world and transform it into something greater and more beautiful.

4.2.2 The creation as means by which man can glorify God

Man's ability to practise science was not only intended to allow him to gain greater control over the earth. It was also intended to bring him to a realisation of the wonder, majesty, power, intransience and glory of the Creator. This he would realise through his research of, among others, the order, greatness, age and forces within the creation.

By saying this, we want to emphasise what has already been stated. Nature and the study of nature is not meant to provide some kind of laboratory in which man himself may hopefully discover God. At best, science can bring us to the realisation that there must be a God. However, this kind of notion of God is but vague and abstract. In actual fact it is little more than a logical deduction. The true God only becomes a reality to us when He graciously reveals Himself to us. Only when the Holy Spirit witnesses in man's spirit that God has made him a participant of his own nature has such a person truly found his Creator.

Through the experience of union with God, man also experiences a unity with the creation as God's handiwork. Man then finds that he exists in harmony with nature (and with himself). In addition he discovers that creation reveals something about God himself. The study of nature becomes like a sounding-board. It causes the peace, reverence, joy, love et cetera experienced by man as an outcome of his relationship with God to take on new dimensions and become more intense. This is what a man, like David, experienced in his observation of, and contact with, nature (cf Ps 18–19). These Psalms are not written to disclose scientific facts as such. Instead they describe how David experiences the wonder of God's work of creation in nature. Although we are not dealing with a rational and scientific view when we read the words of David, they nevertheless impart a truth and an insight which scientific research cannot offer. This is a case where you are enabled by God's Spirit to become aware of the Creator of all these things through your observation of creation. In addition you also become aware that the meaning and purpose of all that you observe is to glorify God. When you observe these things you are touched in your heart by the Lord with the result that you cannot but delight yourself in his handiwork.

4.2.3 Man as social being from the perspective of creation

Through the creation God also ensured that man would exist and develop as a social being. We see how the Lord God created people as husband and as wife with the mandate to: "Be fruitful and multiply; fill the earth ..." (Gen 1:28). Besides man's relationship to nature, he also has to live in a relationship with his neighbour. Among others, social interaction is necessary to develop man's moral characteristics. We refer to love, goodness, mercy, self-control, perseverance and other similar virtues which build a good and sound character. Together with his neighbour, man does not only practise science and technology but he also creates culture whereby he is able to produce all beautiful and good things.

4.2.4 The Fall from the perspective of creation

We know how the Fall disrupted the relationship between man and God, between man and his fellowman and between man and nature. The outcome of this is death and all the grief which man experiences daily. However God's plan and purpose with man has not thus been thwarted. Although God allowed evil and sin to occur, these anti-God forces cannot hinder the unfolding of his plan and purpose for man. As we have already seen in earlier chapters (those dealing with hamartology), evil and every sinful event has to contribute to the realisation of God's plan and purpose with everything. The covenant of grace came about because of the Fall. The Fall necessitated Christ's coming to this earth where He would again make everything new and good by his work of re-creation – something that will only be completed at the end of the world.

The re-creation work of Christ added a very special dimension to the present creation. In the re-creation of that which had already been created, the fallen creation itself also obtained meaning. The meaning lies therein that man would have a particular history with the Lord. It is a history in which God's saving grace and love in the midst of sin, his faithfulness and care under trying circumstances, his power under the attack of evil forces and his victory in spite of corruption and death would be manifested. In this way man could experience and become that which Adam and Eve could never experience before their fall.

In summary we can say that through the creation of everything, but in particular of man, God wanted to share his glory with those different to himself. For this reason the creation tells of God's eternal love.

4.3 The significance and purpose of the creation with a view to God, the Father

When we refer to the significance and purpose of creation with a view to God, we do not want to suggest that we have insight into God's eternal counsel. All we want to deal with is that which we as mere mortals can deduce from God's word and handiwork.

4.3.1 Creation confirms that God is the Completely Other One

Through the creation God not only brought forth that which is different from Himself, but also that which, in some measure, is removed from Himself. By means of this God gave all created things an own space together with laws unique to their own nature and function. Furthermore, He gave morally responsible beings (angels and people) the right to self-determination, although this can only operate within certain parameters. This own will and self-determination, among others, implies the freedom to obey or disobey God. In this way God allowed these beings to manage their lives according to their own insight and will. In this way they also become responsible for their eventual destiny.

In this regard we reject the idea that God needs creation in order to stand in a relationship to that which is different to Himself. God, in his perfection, does not need this kind of relationship. In the eternal Trinity, He already stands in a relationship to the Son and the Holy Spirit.

It is for the very reason that God did not need the creation in order to be complete in Himself that the world is not divine. Everything on this earth functions according to its own order and contingence, as has already been mentioned. If we do not keep this in mind, then creation could be seen as an extension of God through Christ. Then we lapse into pantheism.[13]

On the other hand we should not only point out God's transcendence with regard to the creation, but also his immanence. He still maintains everything. He knows about everything that happens and he guides everything to a determined destiny. If we do not view it in this way, then we lapse into deism.

Created things did not only originate through the invisible (the divine), but also stand moment by moment in an indestructible relationship with God. Without Him nothing can exist.

4.3.2 Creation and God's love and righteousness

The Lord has determined that those saved through Jesus Christ will be heirs of God and co-heirs with Christ (Rom 8:16–17). This confirms God's righteousness (1 John 1:9; Rom 3:21–26). Those who reject Christ will suffer God's righteous judgement. This will also confirm God's justice (Rom 3:5).

4.3.3 Creation and the glorification of God

As has already been mentioned, God has also determined that all enemies will be subjected beneath the feet of Christ. When everything has been subjected to Christ, He, together with the host of the redeemed, will subject Himself to God, the Father. Thus God will be all in all (1 Cor 15:25–28). This means that everything which took place in creation will ultimately be realised to the glory of God. Consequently the meaning and purpose of creation will reach its climax.

By this final act when God will become all in all, his immanence or presence in the life of man will be revealed as never before. Thus the situation (with its accompanying risk) whereby every morally-responsible person had to choose between good and evil is filled by that which is according to God's will. Thereby the history of creation is wound up. Now humanity becomes part of God's eternity where He deals with everyone directly and per-

13 Colin E Gunton points this out: "It is the relation in otherness between God and the world that is conceived with the help of the doctrine of the Trinity, and probably cannot adequately be conceived in any other way." Op cit, p 147.

fectly whether in his glory or in his wrath. With regard to the believer, this is realised in the new heavens and the new earth where God's righteousness will reign. After all *this* is God's final purpose with this creation!

CHAPTER 4 Evolutionism

Evolutionism is a subject which does not really belong in dogmatics. However, on account of its connection to protology it merits attention. Evolutionism stands in contrast to the biblical doctrine of creation, and is very often commonly embraced. Therefore we need to account for it from a theological perspective.

By evolutionism we do not mean the theory of evolution as such but the philosophy built around it. We speak of evolutionism when, on the basis of the theory of evolution, faith in God, the Creator, is interpreted as something unscientific which should be rejected. Instead of faith in a Creator the evolutionist (the term evolutionist is used in this connection to indicate one who teaches evolutionism) propounds that everything which exists, developed spontaneously and accidentally out of matter. This also includes all forms of life, yes, even man.

Therefore, when referring to evolutionism we are not only dealing with materialism but more especially with blatant atheism. As a rule evolutionism is the faith espoused by the atheist. It even rejects the idea that the principle of evolutionism or development from a lower to a higher form of life can be a method used by God as Creator. According to the atheist, no Creator-God, who willed, planned and brought forth everything which we observe, exists. According to him matter is the only thing that exists and the only reality that there is.

According to evolutionism matter always existed, that is, it is something which is eternal in itself. By chance and selection (something which also took place coincidentally) matter brought forth the things which we observe today, yes it even brought forth us, those who observe.

In the discussion of the meaning and implications of evolutionism, we shall highlight certain points of doctrinal interest.

1 EVOLUTIONISM IS BASED ON A THEORY

The idea of evolution used as a basis and point of departure for evolutionism, merely remains a theory. It is a hypothesis according to which the scientist endeavours to explain certain natural phenomena.

Whilst there are many things which ostensibly point to evolution, there is, however, still insufficient data to present it as a scientifically proven truth. In the process of evolution from the most simple cell to the most complex organism, there are still many questions which cannot be answered satisfactorily. Many events are still relegated to coincidences. Therefore it is little wonder that a growing number of scientists have misgivings about the way in which evolution is presented.[1]

In this connection there are many examples indicating that evolution is only a theory. Let us consider a number of these.

First and foremost we refer to the idea that one species was to have developed into another more complex one.

It has been scientifically proven that, because of a variety of reasons, mutations and changes can take place within a species. Whether one species can change into another species is something that cannot yet be scientifically observed or proven. Such a notion is merely hypothesised. However, this links up with the explanation postulated by evolution for the existence of different species from the most simple to the most complex organism. No *links* exists which can undeniably prove that one species can – whether by coincidence or supernatural design – develop into another species.

Moreover no *father species* have been found from which different species developed. This applies in particular to the link between man and apes proposed by evolutionists. Paleontology does refer to the existence of ape-like people (pre-hominids), yet nothing could be found which indicates a common primeval ancestor of man and ape.[2]

Darwinism explains that similarities among living beings indicate a certain common ancestor. However, the fact is that many examples prove this supposition false. Similarities between beings of different species exist which, scientifically spoken, have no connection with one another.[3]

Evolutionists maintain that the human embryo demonstrates the different stages of evolution from fish to man. Thus it is alleged that gills like those of a fish are also found in the human embryo. However, scientific research

1 Dr Francis Crick (Nobel prize winner together with James Watson with regard to the explanation of the DNA-structure) writes: "The theory (of evolution) seems plausible, but as a theory it suffers from a major defect: it is too accommodating. In a loose sort of way it can explain anything. A second disadvantage is that the early steps needed to get the system going seem to require rather a lot of chance effect. A theory of this sort is not necessarily useless if one can get at the facts experimentally. Unfortunately in this problem this is just what it is so difficult to do." The Origin of the Genetic Code, *Journal of Molecular Biology*, Vol 38, p 367.

"Dr Paul Brand, world renowned surgeon, Chief of the Rehabilitation Branch of the U S Public Health Service, and professor of Surgery, has the following to say about the genes' chemical code DNA that controls all forms of life, maintaining the species propagating each after its own kind according to Genesis chapter one, and preventing the evolution of any species into another: What moves the cells to work together? The secret to membership lies locked away inside each cell nucleus, chemically coiled in a strand of DNA. Once the egg and sperm share their inheritance, the DNA chemical ladder splits down the center of every gene much as the teeth of a zipper pull apart. DNA reforms itself each time the cell divides: 2, 4, 8, 16, 32 cells, each with the identical DNA. Along the way cells specialize, but carries the entire instruction book of one hundred thousand genes'." Cf Duffield and Van Cleave, *Foundations of Pentecostal Theology*, p 119.

2 J J Duyvené de Wit has the following to say about this: "Between the presupposed ancestor of the oldest known fossil man and the most human-like animal primate that could be considered his nearest cousin from the transformistic point of view, lies a gap of over forty million years in which no connecting links whatsoever between animal and man have been found." *The Palaeontological Record and the Origin of Man*, 1963.

3 Wilder-Smith quotes instances where similarity has nothing to do with genetic kinship, like the similarity between the eye of an octopus and that of a human, the pouches of marsupial mice which resemble those of marsupial bears. Op cit, p 53.

has proven that that which has the appearance of gills, bears no similarity to the gills of a fish. Study of the embryos of birds shows no signs of teeth as found in a few fossils of extinct flying animals, which, according to evolutionists, were the ancestors of our birds.[4]

Up to the present time, there exists no real scientific evidence of evolutionary transitions. Darwin himself conceded that the geological record of the time when he wrote his book on evolution (1859), was extremely imperfect. He says, *inter alia*: "... this fact will to a large extent explain why we do not find intermediate varieties, connecting together all the extinct and existing forms of life by the finest graduated steps. He who rejects these views on the nature of the geological record, will rightly reject my whole theory."[5] What do the geological and archaeological findings up to the present day reveal? About this Luther D Sunderland remarks: "Now, after over 120 years of the most extensive and painstaking geological exploration of every continent and ocean bottom, the picture is infinitely more vivid and complete than it was in 1859. Formations have been discovered containing hundreds of billions of fossils and our museums are now filled with over 100 million fossils of 250 000 species ... With every geological formation explored and every fossil classified, it has become apparent that these, the only direct scientific evidences relating to the history of life, still do not provide any evidence for which Darwin so fervently longed. The gaps between major groups of organisms have been growing even wider and more undeniable."[6]

Michael Denton points out that there was a notion that through increasing the scope and intensity of the sampling of fossils the gaps would narrow. That did not happen and the gaps are as intense as ever. "[A]lmost every major advance in biological knowledge, from the founding of comparative anatomy and paleontology in the eighteenth century to the recent discoveries of molecular biology, has only tended to emphasize the depth and profundity of the great divisions of nature."[7] This statement has been confirmed by several renowned scientists.[8]

The notion that lower levels of organisms had developed into more complicated ones, is seriously questioned by scientific data. "The first appearance of fossils (Cambrian, 600 million years ago by the fossil reckoning) was at that time in a pretty highly developed form. They don't start very low and evolve bit by bit over long periods of time. In the lowest fossil-bearing strata of all, they are already there and are pretty complicated in more-or-less modern form ... Everything was there at the opening of the drama. The curtain goes up and you have the players on the stage already, entirely in modern costumes ... thus no sign of slow build-up ... No intermediate forms

4 Cf S Maxwell Coder and George F Howe, *The Bible, Science and Creation*, 1966, p 71.

5 Charles Darwin, *The Origin of Species by means of Natural Selection.* 6th ed, 1902, pp 341–342.

6 Luther D Sunderland, *Darwin's Enigma. Revised ed*, 1988, p 9.

7 Michael Denton, *Evolution, A theory in crisis*, 1986, p 353.

8 Cf David Raup, curator of geology at the Field Museum of Natural History in Chicago; Colin Patterson, a senior paleontologist and editor of a prestigious journal at the British Museum of Natural History; evolutionists like Stephen Jay Gould and Isaac Asimov. Cf Luther D Sunderland, op cit, pp 10, 11.

that would indicate the common ancestry of life ... but abrupt appearance of all major groups of living organisms."[9]

This fact is also substantiated by the so-called living fossils. Plants and animals which should have been extinct millions of years ago, still exist in the same form today. We mention but one example, namely the coelacanth, a fish in the waters of the East African coast. The living specimen is just the same as its fossil, which was thought to have been extinct for one hundred million years.

Fossils of all kinds of plants, fish, large amphibians, insects and reptiles occur in the Cambrian stratum, whilst the period prior to the Cambrian stratum shows little signs of living beings. From which ancestry did these plants and animals then develop? It appears that all the plants and animals mentioned must have originated in one period as the Cambrian stratum indicates. Geologists are so dumbfounded by this phenomenon of an explosion of life that they refer to it as the *Cambrian Explosion.*[10]

The sequence in which certain apelike people would have appeared has already been proved wrong on more than one occasion. In 1974 Donald Johanson discovered the fossilised skeleton of a homo (human) in the Afar region of Ethiopia. The time when *Lucy* (the nickname for this *homo*) lived, is calculated as 3,75 million years ago.[11] It is millions of years older than the apelike people (Neanderthal man, the Cromagnon cave dwellers, the Java man and others) who have always been considered the ancestors of contemporary man. If the idea of evolution was correct, it should have been the other way around here, Lucy should have been millions of years younger. It is therefore no wonder that Johanson became involved in a huge dispute with another paleontologist, L S B Leakey. The latter discovered the *homo habilis* skeleton in Tanzania in 1959.[12]

The so-called family tree portraying the development of man from an animal-like ancestor, is to be doubted very much. The most vital parts of the tree have very little to do with real scientific data, whereas a great deal is based on unconfirmed assumptions.[13]

Footprints of the Brontosaurus, as well as footprints that concur with those of modern man, exist in the bed of the Paluxy River, Glen Rose, Texas. According to the theory of evolutionism, prehistoric animals from the Cretaceous period (130–200 million years ago) lived while the prehominid and modern man appeared much more recently.[14]

It should always be kept in mind that the portrayals of prehistoric animals and hominids rest on scant evidence. In the case of hominids everything is based on a skull or a part thereof, a jawbone with a few teeth, a femur or a few parts of a skeleton. In addition there are sometimes a few imple-

9 Cf Luther D Sunderland, op cit, pp 150, 151.
10 Cf S Maxwell Coder and George F Howe, op cit, pp 64, 65.
11 Cf Lucy, *The beginnings of humankind* by Donald Johanson and Maitland Edey, abridged for the Reader's Digest. Nov 1981.
12 Cf L S B Leakey, Finding the World's Earliest Man, *National Geographic*, 118(3), 1960, pp 420–435.
13 Luther D Sunderland reports: "Richard Leakey summed up the situation on the final Walter Cronkite 'Universe' program. He said that if he were going to draw a family tree for man, he would just draw a huge question mark." Op cit, p 87.
14 Cf A E Wilder-Smith, op cit, pp 135–142.

ments. To derive the appearance of an apelike man from these: the soft parts of the body, the skin with its hair, the measure in which he had a stooped or an erect position, as well as his intellectual abilities and so on, demands considerable imagination.

The determination of the age of a fossil (referring to the time when it was still part of a living being) takes place by measuring the radio-activity or the so-called carbon (C14) test. However, it appears that reasonably reliable calculations can only be made in this way when cases of 30 000 years or less are dealt with.[15]

Owing to the many flaws in Darwinism, it is no wonder that evolutionists, such as G A Kerkut, declare that the general evolutionary theory and the proofs which substantiate it are not cogent enough to make it any more than a working hypothesis.[16]

2 IT IS A VIEW UNSUBSTANTIATED BY THE NATURAL SCIENCES

First and foremost we point out that there are no scientifically substantiated grounds for deducing that matter either originated from nothing or is itself eternal. From a scientific point of view it is nonsense to say that something can develop from nothing.

Moreover it is just as unscientific to postulate that matter is eternal. Eternal means that all time and space is transcended, whilst science can only function within time and space. Without these dimensions there is nothing which can be researched scientifically. One who speaks of something *eternal* can no longer do so on the grounds of the findings of the natural sciences. It can only be done on the grounds of theological or metaphysical considerations. In this case one is in the area where one has to speak either on the basis of what is revealed to him from Above or what one postulates for oneself. When believing in something which transcends human observation, one inevitably moves into the area of religion.

For this reason we can say that evolutionism is also a kind of religion. Instead of believing in God, one believes in matter. In this respect we point out that the Bible considers any denial of God's existence to be the pronouncement of a fool (Ps 14:1).

3 HOW EVOLUTIONISM DEIFIES MATTER

According to the evolutionist, not only is matter the only thing which exists, but characteristics unique to God are ascribed to it. Evolutionism maintains or supposes that matter was never brought forth by a higher reality but that it, being itself eternal, brought forth all other things.

Furthermore it supposes that matter possesses powers and possibilities which can only be described as divine. It is able to bring forth other things which transcend all human thinking. It is responsible for the coming into being of the macro-cosmos as well as the micro-cosmos. It is able to devel-

15 Cf A E Wilder-Smith, op cit, pp 116, 117, as well as Coder and Howe, op cit, p 79. They speak of 5 000 years. According to the most recent methods, it is claimed that it may be reliable up to 30 000 years.
16 G A Kerkut, *The Implications of Evolution*, pp 155–157.

op structures and systems (such as the heavenlies, the atom, the human body) which are so complicated but also witness to perfect design and balance, with the result that man is overwhelmed with wonderment. On the biogenetic level it is able to cause life to develop out of that which is lifeless. Furthermore, it is able to cause a living cell to develop into the most complicated organisms and then sustain them for years. It is even able to bring forth living creatures such as man – a being with self-consciousness, volition, cognition, memory, emotions, desires, imagination, responsibility, aesthetic aptitude and an appreciation for the beautiful and the good, business acumen, the ability to practise science, yes, even the ability to postulate something such as evolutionism whereby he thinks he can determine his own origin and significance!

Just because we cannot observe and monitor God in the same way as other natural phenomenon, some people reject his existence. Thereby they also reject the thought that He created everything. (Imagine if because you could not find the carpenter in the wood of a table made by him, you deduced that no such person exists!). Merely because such people know nothing about God's merciful revelation in a person's life, they create for themselves their own god whom they can observe and monitor. Instead of belief in God, the Creator and Sustainer of everything which exists, the evolutionist exalts matter to deity. Where the Bible teaches that man was created according to the image and likeness of God (Gen 1:26),evolutionism swaps it around. It teaches that *God* should be represented according to the image which man has derived from evolutionism. Therefore we can also regard evolutionism as a form of idolatry, in that material things are exalted to gods (Rom 1:22, 23).[17]

4 IT IS A PREJUDICED AND ARROGANT POINT OF VIEW

We also detect how prejudiced and arrogant the viewpoint of the evolutionist is. On the one hand, he rejects in advance any knowledge which he cannot obtain and evaluate according to normal scientific procedure. On the other hand, he does not hesitate with his evolutionism to cross the boundary into the area of eternal realities. He regards himself capable of explaining the impenetrable mystery about eternity and the origin, meaning and destiny of everything he observes! Instead of humbly acknowledging that no human thinking nor scientific practice can cross the border to the eternal Source and Sustainer (God) of all, the evolutionist does just that! He decides that he does not need God and that he has the answer for the origin and the meaning of everything! How almighty and all-knowing man can be in his own eyes! Not only does he show an arrogant superiority, but he also taunts God (Ps 2:1–4).

By contrast, the true scientist is aware of the parameters within which he should practise science. He knows that the observable things which he studies and which exist within time and space, had to be included and

17 C F Phillip E Johnson has the following to say about this matter: "Evolution holds the promise of ultimate heaven and bliss, because the plurality of individual thoughts combine and mutually reinforce each other in a single act of unanimous Thought ... a harmonized collectivity of consciousness equivalent to a kind of super consciousness." *Darwin on Trial*, 1991, p 130.

brought forth by that which transcends all these things. He also knows that he, as a mere man, cannot cross the border into this eternal reality and discover it by himself. This knowledge can only be obtained in a supra-scientific way, or as the Bible puts it – through God's revelation to man.

Furthermore, an essential scientific principle is to be receptive to any form of knowledge available to man. If there are millions of people on earth – including many respected scientists – who testify that God has revealed himself to them through the Bible and his Spirit, then one should heed this. Every scientist has to ask himself whether he cannot also arrive at a revelation of and an encounter with God, even though in a supra-scientific way! Scientists who testify to these things, find that this is not contradictory to true science. On the contrary, it provides the necessary frame of reference within which they practise their science and give it meaning and significance.

5 THE EVOLUTIONIST HAS A TOTALLY UNACCEPTABLE IDEA OF WHAT MAN ACTUALLY IS

The evolutionist regards man as a mere organism which developed through an evolutionary process from a single cell. This cell merely originated spontaneously from lifeless things. This implies that man's existence is only coincidental. He is there without any agent wanting him to exist and without a reason for his existence. Therefore his life has no sense or destination since he is only the product of a blind, meaningless, accidental and material evolutionary process!

The evolutionist does not acknowledge the reality of man as soul or spirit which is completely different from his physical body. According to him man is merely a thing (albeit a very complex one) alongside other things. To him self-revelations such as self-knowledge, self-consciousness, own will, emotions, thinking, passions et cetera are only phenomena which matter attains by a coincidental evolutionary process.

In reality it amounts to this: that man has to deny himself as a subject. Though man as subject reveals himself through material substance, he as a self-conscious and self-willed being, is totally different from mere matter, such as a piece of rock. Yet the evolutionist wants man to say to himself: "I am telling you that you are not an *I*; you are merely an *it*, a material robot which incidentally developed in this way." One inevitably wonders who and what that phenomenon is who can distance himself from himself, by describing himself as mere matter! What we face here is an illogical and senseless short-circuit which man has within himself. It is a ridiculous endeavour to try and get rid of himself as the subject – a being who wants to rid himself of himself!

One wonders if the evolutionist would like others to treat him according to what he says he is, a mere thing and a robot which per chance came into existence! If he is only a robot, he is merely a thing which others can either use as they please, or cast away as useless, or reduce to pieces in an attempt to put a better one together, or burn up as something superfluous!

On the other hand, the believer knows (and I should imagine everyone who thinks soundly) that he, as self-conscious and self-willing subject, is not matter such as a stone; but that he reveals himself through physical things, his brain, nervous system and his entire body.

He believes that there is One (God) who transcends man and all visible things. He believes that this One created and sustains all things. He knows that he as a subject (an I) was brought forth by an infinitely greater and eternal Subject (the I). With this knowledge he can find peace. In this he is different from those who find themselves living in the darkness of evolutionism where they try and believe contradictory things about themselves – things without beginning, without purpose and without meaning.

6 THE EVOLUTIONIST BELIEVES IN GREATER MIRACLES THAN THOSE MENTIONED IN THE BIBLE

The evolutionist (together with certain theologians) does not believe in the miracles which we read about in the Bible. These people think they can explain why simple people still talk about and believe in miracles. Yet, on the other hand, the evolutionist makes statements which presuppose faith in greater miracles than those read of in the Bible. We look at a few examples.

The evolutionist wants us to accept that matter had no origin and was always there. Evolutionism maintains that lifeless things can bring forth life.

Furthermore, it declares that molecules were shaken by a force (which must have accidentally been there) in such a way that over a long period they accidentally ended up in a configuration from which the first living cell originated. On the grounds of the laws of chemistry and physics – which also originated by chance – more complex organisms emanated from this first living cell. These developed into plants, sea animals, amphibians, reptiles, birds, mammals and eventually man.

This evolutionist view can be presented concretely by an example. A space ship which has to take people from earth to the moon and back has just been launched. Somebody steps forward and tells those responsible for the building and the launching of the spaceship: "This ship, as it has been constructed from different materials together with all its computers and intricate instruments, could come about without any human agency in that over a very long period molecules can group themselves together by chance in this way." It goes without saying that those who hear this will think that the speaker is out of his mind. But let us now pay attention to what the evolutionist says about man. Man with his wonderful body and psychological abilities is unthinkably more complex than anything which he can build. The evolutionist who has to acknowledge that the spaceship has to have a maker in the form of man, does not believe that man has a maker! Man originated by chance and that by an accidental grouping together of molecules which also accidentally originated! Who can deny that an evolutionist believes in miracles!

All that can still be said about this is that people are prepared to even betray their sound minds as long as they can eliminate God![18] How terrible is the influence of anti-God and evil forces in people's lives!

Let us, as a matter of interest, pay attention to this *chance* which some

18 L T More, late professor at the University of Cincinnati, confessed in his book *The Dogma of Evolution* the following: "Our faith in the idea of evolution depends upon our reluctance to accept the antagonistic doctrine of special creation." *What the Scientists say about Evolution, Dedication*, Vol 9, no 2, p 7.

people have made a creator of everything. Foremost chance means that things can develop in any direction. Particles of matter may group together in a unit or scatter. When one tosses a coin several times, there will be an equal occurrence of heads and tails. The longer the process continues, the greater the possibility that there will be a head for every tail. This is after all a mathematical and a statistical axiom.

The same principle is also valid for the integration and disintegration of particles of matter. For every accidental integration there is also a disintegration to cancel it again. To declare that only one coincidence, namely integration, happened exclusively or predominantly so that an upward evolutionary process could be maintained, implies nothing less than a miracle, that is, the activation of another factor at work outside of and above mere chance.

In reality not everything is improving or evolving in a positive way. According to a law of thermodynamics, there is an overwhelmingly great possibility that everything is moving towards eventual collapse. This applies to the stars and the entire universe. On earth the mountains are being weathered; iron rusts; living creatures become weak, sickly and eventually die. Even the temporary life which appears whenever reproduction takes place is at the cost of the energy radiated by the sun. Eventually the sun will become so weak that life on earth will not be possible.[19]

7 THE EVOLUTIONIST'S IDEA THAT LIFE ORIGINATED IN A COINCIDENTAL WAY IS TOTALLY IMPROBABLE FROM A MATHEMATICAL PERSPECTIVE

So many things must be precisely adjusted to make life on earth possible that the possibility of this happening by chance and on its own can be discounted.

7.1 The fine tuning regarding the entire universe

In the first instance we refer to the amazing *fine tuning* of everything in the universe in order to produce a planet like the earth with its special features which make life on it possible. We look at a number of examples.

Recently it was discovered that after the Big Bang little waves or fluctuations in the radiation of plasma occurred which was the reason for the formation of galaxies. As a rule radiation is extremely uniform wherever it may happen, but with the origin of the universe it was different.[20]

To produce solid matter – something necessary for life – there had to be

19 "Barrow and Tipler estimate that the length of time that the earth will still contain a biosphere, and thus remain suitable as a habitable planet, may only be another $3\frac{1}{2}$ minutes in a time scale where the remaining lifetime of our sun, approximately 4,5 billion years, equals 365 days. One minute on this time scale corresponds to 8 562 years; $3\frac{1}{2}$ minutes is equivalent to 30 000 years." David L. Block, op cit, pp 24, 25.

20 Stephen Hawking, Lucasian Professor of Mathematics at the University of Cambridge (the same chair once held by Sir Isaac Newton) considers this as "the discovery of the century, if not all time."

George Smoot, team leader of COBE (Cosmic Background Explorer satellite launched by NASA in November 1989) remarked: "What we have found is evidence for the birth of the universe. It's like looking at God." Quoted by David L Block, op cit, p 10.

the right balance between particles and anti-particles. For every 100 million anti-protons in the early universe, there had to be one more proton.[21] The density of the mass (plasma) destined to become galaxies must be precisely balanced with the speed with which the universe expands. If the speed of the expansion was just a millionth of a millionth slower, the entire universe would have collapsed. Instead of stars there would have been only so-called *black holes.*[22] Without stars there would also not have been an earth nor people on it.

The forces which bind the universe together are: (a) The electrical or electromagnetic force which controls the structure and interaction of atoms and molecules; (b) the gravitational force which is responsible for the structure and dynamics of galaxies, stars and planets; (c) a weak nuclear force which operates on atomic particles over very short ranges; and (d) a strong nuclear force which holds the protons in the atomic nucleus together. This is necessary because protons are positively charged and will repel one another if there is no force to stop this. These forces must all function in perfect balance. If for instance the electromagnetic force was stronger in its relation to gravity, there would have only been cold red stars. There would have been no supernovae; and the explosion of such new stars is absolutely essential for the formation of carbon which is again necessary for life. If the electromagnetic power was slightly weaker, all stars would have been warm and blue. As a result the lifespan of stars would have been too short and their heat too great to make the origin of life possible.[23]

The universe must actually be 15 000 million years old and it has therefore expanded to this extent in order to bring about the necessary cooling which makes life on earth possible. A younger and smaller universe would mean that there could be no life on earth.[24]

7.2 Fine tuning with regard to the earth

The earth's position in the solar system and its characteristics have to be precisely tuned in order to make life on this planet possible. Here are a few examples.

The earth must be the right size in relation to the sun. The earth has to be the right distance from the sun. The earth has to revolve around the sun in a certain time so that the seasons are neither too long nor too short. A summer or a winter lasting years respectively would create such extreme climatic conditions that life on earth would be impossible. The speed with which the earth revolves on its axis and which determines the duration of day and night must also be tuned correctly. If the periods of day and night

21 David L Block remarks: "The 100 million anti-protons annihilate with 100 million protons at very early times, and the one remaining proton becomes part of the matter-dominated universe we live in today! ... If the ratio of protons had exactly equalled the ratio of anti-protons, all matter and antimatter would have annihilated within 1/100 000th of a second to produce a universe of pure light. And we would not exist." Op cit, p 29.
22 Cf David L Block, op cit, p 12.
23 Cf David L Block, op cit, pp 14, 15.
24 As a result of the discoveries of cosmologists John Barrow and Frank Tipler, David L Block remarks: "We can therefore say that no one should be surprised that the universe is so large. We could not exist in one that was any smaller." Op cit, p 16.

each lasted for weeks or longer, there would be such extremes of heat and cold on earth that life as we know it would not have been able to exist. The earth's mass must be just right so that its gravity can allow the normal, necessary mobility of living beings. An atmosphere with its specific density and composition is essential, otherwise life on earth would also be impossible. The atmosphere ensures that meteorites attracted by the earth burn out timeously; that there is the necessary protection against dangerous rays of the sun (the function of the ozone layer); that clouds form so that there is rain; that there is the necessary oxygen for man, while ensuring the necessary carbon dioxide for the plants, and so on. The oceans are needed to create a favourable climate for the living beings on dry ground; the balance between plant and animal life is necessary because they are mutually dependent. Basic materials on earth are necessary for living beings to develop. Moreover, there are many other aspects of the earth which have to be tuned correctly to make life possible on earth.

7.3 The tuning in connection with biochemical matters

Besides correct astrophysical and geophysical tunings, all biochemical factors must also be precisely tuned to make life on earth possible.

Without entering into too much detail, we mention a few examples expounded by David Block.[25]

We know that the DNA chain with its 110 000 different genes is the blueprint for the development of an organism such as man with regard to his psychosomatic existence. Once again every gene has between 180 and 360 nucleotides, whilst each one of these in the DNA chain represents one of the four *letters:* A, C, G and T. The *words* constituted when the four letters are arranged in their proper order are genes.

In connection with this let us now look at certain mathematical realities.

Deley (Evolutionary Biology) calculates that the chances of a coincidental production of a certain gene with the required characteristic lies between one in $4^{180} = 2{,}3 \times 10^{180}$ and $4^{360} = 5{,}5 \times 10^{216}$. In such a gene the letters A C G T should occur in the correct sequence of its nucleotide. John Barrow and Frank Tipler (The Anthropic Cosmological Principle) confirm the above observation. As a result they calculate that the 4,5 billion years of the earth's existence is far too short a period to make the accidental development of man possible. Besides those mentioned by David Block, there are also several other prominent scientists who have come to the same conclusion. Douglas Hofstadler maintains: "There are various theories on the origin of life. They all run aground on this most central question: How did the Genetic Code, along with the mechanisms for its translation (ribosomes and R.N.A. molecules), originate? For the moment, we will have to content ourselves with a sense of wonder and awe, rather than with an answer." Eigen argues: "Which came first, function or information? As we shall show, neither one could precede the other; they had to evolve together."[26]

25 Cf David L Block, op cit, pp 17, 22.
26 Cf Luther D Sunderland, op cit, pp 145, 146.

Furthermore, at an estimate the human genome contains 110 000 different genes, of which it is unlikely that every key gene can appear more than once in 15 000 million years in a coincidental way. When we come to the chances of a coincidental assembling by which man (homo sapiens) can evolve, the possibility lies between one out of

$4^{180(110\,000)} = 10^{12x10^6}$

and one of

$4^{360(110\,000)} = 10^{24x10^6}$.

If we convert the minimum number

10^{12x10^6}

in terms of millions, then the possibility becomes one out of

$1\,000\,000^{119999994}$.

If we want to express verbally the latter figure in terms of millions by saying: "times a million" every two seconds for twenty four hours everyday, we would have to speak without interruption for nine months and one week to complete this figure. It is one chance in this incomprehensibly large number which makes it possible for man to have originated accidentally. And it is on this immensely huge improbability that evolutionism is based. If we want to be consistent according to the accidental notion, then we must say that the things relevant to this calculation also had to come about accidentally. Because it is an undisputed fact that nothing can come forth from nothing, not even this one possibility in that completely incomprehensible number is valid! How does the idea of coincidence of evolutionism measure up to this?

7.4 The possibility of life on another planet

Since the universe is so immensely big (15 billion light years in diameter), it is alleged that there must be a star like our sun, with a planet precisely with the same characteristics as our earth, somewhere in this vast universe. Such a planet or planets would then contain living beings comparable to man.

To calculate this possibility mathematically, we again refer to the exposition of David Block.[27]

> 'A realistic estimate of the number of galaxies in our visible universe is conservatively estimated 10^9; let us accept 10^{11}. Let us further suppose that every galaxy has the same number of stars as our Milky Way (10^{11}). This gives us a total number of stars of 10^{22} within the present day observable horizon. Even if the possibility for the development of intelligent life on a planet would be as high as one in 10^{30} (in reality it is less), then there are only but 10^{22} stars as candidates!'

Therefore the possibility of living beings comparable to the people here on earth existing somewhere on another planet is totally improbable.

7.5 General conclusion

All the fine tunings in the universe which have been referred to, not only exclude all coincidence, but definitely indicate that everything exists accord-

27 Cf David L Block, op cit p 23.

ing to an underlying plan and that a supernatural agent must have accomplished this. Even those who advocate a biological evolution come to the conclusion that there must be a divine providence according to which everything exists and operates.[28]

Scientists arrive at the so-called anthropic principle concerning the origin of life. This means that everything in the universe actually took place in a predesigned way, in order for living beings (especially man) to appear – the so-called anthropic principle (Greek: *anthropos* = man). True, there are some scientists who are not prepared to draw the same conclusion.[29] Yet there are also certain eminent astronomers and astrophysicists who see no other explanation but that there must be a God who has determined everything so that we as people could be here.[30]

This all serves as proof that the so-called scientific grounds on which evolutionism bases its view has no validity.

8 EVOLUTIONISM IMPLIES THE DESTRUCTION OF MORALITY

Because the evolutionist thinks only in materialistic terms and therefore also atheistically about man, he simultaneously rejects any human responsibility and accountability. Whatever man is and does is merely the result of accidental evolution where everything is determined mechanistically. Since man cannot be regarded as a self-conscious and morally-responsible subject distinct from mere matter, there can be no mention of morality. Thus man no longer stands as subject in a responsible relationship to God, *the* Subject. No longer can man be responsible for what he does to himself, his neighbour or nature. The consistent application of this kind of reasoning means the total downfall of society.

According to the evolutionistic principle of the survival of the fittest, every person is programmed or tuned merely to do what is in the interest of his own progress or evolution. Moral values such as honesty, compassion for others, a sense of responsibility, self-respect, respect and love for one's neighbour and so forth, are no longer of any consequence. All that matters is that which can facilitate one's upward evolution.

Should this upward evolution demand that people be deceived, assaulted, robbed, mistreated and even killed, then this is right! Should the stronger individual destroy the weaker or the stronger group oppress the weaker or annihilate them, it is in order! After all it concerns the survival of the fittest!

We see something of this evolutionism in national-socialism at the time when Hitler was dictator of Germany. According to this view the Germans as the so-called *Herrenvolk* (superior nation) had the right to annihilate those who were regarded to be from a lower strata (in this case the Jews). No

28 That is the finding of Arno A Penzias, Nobel prize winner in physics, and of Sir John Eccles, F R S, Nobel prize winner in medicine-physiology. Cf David L Block, op cit, pp 2, 3.

29 Cf John Gribben and Martin Rees, op cit p 287. Cf also Colin E Gunton op cit p 116.

30 David L Block mentions the following: Freeman Dysan (*Disturbing the Universe*), Allan Sandage (*Origins – The lives and worlds of modern cosmologists*), Sir Fred Hoyle (*Religion and the Scientists*) and George Greenstein (*The symbiotic universe: life and mind in the cosmos*). Op cit, pp 29, 30.

faithfulness in marriage and family bonds between parents and children are necessary. All that matters is that *good* men will beget *good* children with *good* girls, who will be able to build a super race.[31]

We also see the outcome of evolutionism and accompanying atheism in communism as propagated by Marx, Engels, Lenin and their followers. Faith and dependence on God is regarded as a lapse into weakness. For this reason Marx maintained that religion is like opium to man. All that one has to believe in is the power of reason and science of which evolutionism is a cornerstone. The state is viewed as an evolutionary machine in which the individual is a mere cog. Outside of this function the individual cannot claim any right for himself. If he does not suitably adapt to the state's ideology, he is simply discarded.

When we compare the morality of true Christianity to evolutionism, then it is as light compared to darkness; as freedom in contrast with slavery; as life contrary to death. The Christian believes that God created and sustains everything and guides it to a better destiny. He also professes that man exists in a relationship to God, his Maker, and that he is responsible to him for everything. True virtue is when the perfection of God himself is revealed in a man's life. God is love and the true man is the one who lives a life of love and promotes it. He regards his neighbour as one created in the image and likeness of God. And for this reason he shows a particular respect for his neighbour's life. He endeavours to help him to live a fulfilled life according to God's plan. Every force that tries to fight against God, the Creator, is rejected as sinful and evil. It is also this very force which brings disharmony on the earth, and causes living beings (in particular, man) to suffer and die.

The true relationship with God, as presented by the gospel, is the only hope and salvation for this world. True Christianity combats barbarism, cruelty, slavery, superstition, mistrust between man and man, and so on.

In this connection we again look at the teaching of evolutionism. It alleges that if there is a good God (as propagated by Christians) who created everything and continues to maintain it, then there should not be all the suffering, unrighteousness and death on the earth. They assert that there can be no God precisely because of all these things.

The evolutionist assumes this viewpoint because he rejects biblical teaching. He rejects the view that sin came into this world and marred the good creation. At the same time he rejects the gospel which tells us how God made provision for man to be saved from all this grief. Thus the idea is rejected that God is in control and that good and righteousness will tri-

31 Raymond F Surburg remarks: "In his book *Social Darwinism in American Thought* Prof. Hofstadter has shown that Darwinism was one of the chief sources of racism and of a belligerent ideology. Few people realize that Hitler, in bringing about World War II, merely put into practice what he believed was human evolution. Darwin and Nietzsche were the two philosophers in working out the philosophy set forth in Hitler's *Mein Kampf* ... the blood of the weaker is the life of the stronger." *Darwin, Evolution and Creation*, p 196.

umph over all unrighteousness and suffering.[32]

Instead evolutionism maintains that all the misery experienced by man is merely an accidental and transient phenomenon in the evolutionary process. As coincidence and selection in the evolutionary process continue, all this misery and suffering will eventually disappear. Ultimately chaos will bring forth order, unrighteousness bring forth righteousness, suffering bring forth happiness, and death bring forth life! Blind, impersonal, senseless and illogical coincidence is thus our saviour and only hope!

Furthermore, we should notice that it is Christian morality which considers man to be a special being. It gives the individual the occasion to develop his abilities in the areas of the spiritual, culture, science, technology and so forth. Contrary to this we see how the ethic of materialism and evolutionism degrades man, brutalises him, misuses him and casts him into all kinds of misery.[33] The history of immorality, communism and so on provides us with a shocking example of this. To be truly man, one has to know that one has a divine origin, task and destination. This is what the biblical doctrine of creation offers us.

To this we add the assertion: In order to practise science and to employ it to man's benefit, it must be freed from destructive philosophy and ethics which evolutionism would link it to.[34]

32 A E Wilder-Smith compares the miseries of this world with a beautiful cathedral which is partly ruined by bombs. The fact that so much debris is seen however, is no reason to reject the existence of an architect and builder of such a cathedral. Even among the ruins there are still many things that indicate the design and labour of the architect and builder. In addition, Wilder-Smith remarks: "The creation, even though it is undoubtedly mixed now, being full of the rubble of hate, sin, ugliness, and violence, still does show some pools of love, kindness, beauty and virtue, thereby proclaiming (1) that there is an architect behind these pools, small though they be. In fact, the Bible teaches from cover to cover that illness, death, hate and ugliness are all signs and outward manifestations of a state of ruin' which overtook a once-better world. It also teaches that the signs of ruin' are easily distinguished from the pools' of order, love and virtue which still bear testimony to the state of the original edifice." *The paradox of pain*, 1971, pp 48–52.

33 With regard to this, Raymond F Surburg declares: "In the area of ethics, evolution teaches that there are no absolute standards of right and wrong. Its ethics are naturalistic, monistic, relativistic and pragmatic, because evolution excludes an eternal Lawgiver and an eternal code. In the philosophies influenced by Darwinism, morals change with the vicissitudes of social life. Truth and goodness are determinable by social custom and convention." Op cit, pp 202, 203.

34 Meaningful in this respect is the following remark of the scientist, Sir Frederick Catherwood: "The high tide of evolutionism is beginning to ebb. Some most distinguished scientists can be heard to murmur that belief in evolution, like belief in creation, is a matter of faith. Other distin-

CHAPTER 5 Angelology

Not only did God create those things which are visible to us, but He also created invisible things. We include the angels under the last category. Thus we now turn to the biblical teaching about angels *(angelology)*. This includes the holy angels as well as Satan and the evil spirits.

1 INTRODUCTORY REMARKS

Not only is faith in angels (or similar beings) very old, but it is also widespread. As far as can be determined, supernatural beings, such as angels, have always been interwoven with the religion of mankind.

1.1 Angels and mankind's religious history

The belief exists among world religions that supernatural beings influence man and his world. These beings influence man either for better or worse. In other words, man has to contend with good and evil angels or spirits.

Angels are mentioned in ancient Babylon and Persia. Ancient Egyptians arranged their tombs in such a way that angels could be received in them. In the ancient Greek and Roman civilisations demigods (half man [or beast], half god) are encountered. The mythological literature of these civilisations provide us with a series of these demigods.[1] Mount Olympus was the home of the Greek demigods. One of these, Hermes, was a messenger *(angelos)* from the gods to man. The demigods of both Greeks and Romans were involved with virtually man's entire fate such as love, peace, war, beauty, fertility, the harvest, the nether world and so on.

The idea of angels has gripped people's imagination throughout the ages. There are even examples of so-called inscriptions by angels – the alphabet in the language of angels as well as Hebrew of *Ambelain La Kabbale Prolique.*[2] Furthermore, many paintings, engravings, etchings on window panes and murals etc. record angels and their actions.

1 About this, Arno C Gaebelein writes the following: "The mythologies of nearly all ancient nations speak of such beings. Babylonian mythology pictured them as gods who conveyed messages from gods to men. Roman and Greek mythology had its genuu, semigods, fauns, nymphs and naiads, who visited the earth. Hesiod, next to Homer the earliest Greek poet, said: Millions of spiritual creatures walk the earth'. Egypt and Eastern nations believed in such supernatural, unseen creatures. The belief is well nigh universal." *The Angels of God*, 1825, p 14.
2 Cf Gustav Davidson, *A Dictionary of Angels*, 1967, p 335.

Beside what we read about angels in the Bible, they are also mentioned in the apocryphal books (e.g. Enoch) and in the old rabbinical writings. They are even mentioned in the books of the different non-Christian religions. Thus we find in Islam that angels are described as Allah's messengers.[3]

In biblical times, however, there was a difference of opinion about the existence of angels. Among the Jews there was a group named the Sadducees who, unlike the Pharisees, did not believe in angels nor in the resurrection of the dead (Acts 23:8).

1.2 Angels and the Christian religion

Besides what the Bible reveals about angels, various ideas and opinions about angels occurred in the Christian church. In particular, during the 12th and 13th centuries much speculation about angels arose. People argued about how quickly an angel could move from one place to another, how many angels could stand on the point of a needle, the gender of angels and many other such senseless speculations. These ideas are found in the *Summa Theologiae* of Thomas Aquinas, an important theologian of that time.

In that time names were given to large numbers of angels. These names originated in the various groupings of the letters of the Hebrew alphabet, or simply by placing the suffix el to a word which allowed for the manipulation of the word.[4]

Therefore, it is not surprising that in the period known as the *Enlightenment* (17th and 18th century) when rationalism triumphed, all ideas about the supernatural, including ideas about angels, were discarded. This was a time when medieval cosmology and related subjects were rejected as unscientific. Schools of thought and institutions in Western countries became increasingly secularised. The existence of angels was regarded as either pure mythology or something from the world of fantasy or something which spoke of primitive and pre-scientific views or mere superstition (such as the belief in ghosts).

This new thinking also influenced the church. Since then very little has been written or said about angels in the ranks of the church. Those who still believe in the existence of angels regard them as phenomena which only had significance in biblical times but no longer play a role today.

During the second half of the 20th century however, a new interest in the supernatural has arisen. Together with this the topic of angels has received fresh attention.

This new turn with regard to supernatural phenomena can be ascribed to different factors. One such factor is the unidentifiable flying objects which different people have observed at various places on earth. There were Christians who linked the so-called *flying saucers* to Ezekiel 10:9–17. Consequently reports also appeared of people who had contact with creatures from outer space.

This new interest in supernatural beings is definitely accompanied by the revival of the occult, para-psychology and extra-sensory observations. In-

3 Cf *The meaning of the glorious Koran*, pp 210, 311.
4 Cf Gustav Davidson, op cit, pp 20, 21.

creasing attention is being given to things like satanism, demon possession and exorcism. In addition, we should mention a new attraction for the mystical and non-corporeal beings.

All this can be seen as a reaction to the prevailing rationalism and materialism. People experience the contemporary domination of science and technology as a threat to their spiritual lives in general.

Besides, more and more people today maintain that angels have appeared to them.[5]

1.3 Is there a place for angelology in dogmatics?

Before we answer this question we have to consider the concept "dogmatics" once again. In principle, dogmatics embraces the following points:

It is based on truths revealed by God through Holy Scripture; these truths are revealed in a believer's mind, heart and throughout his entire life through the gracious working of the Holy Spirit, with the result that they obtain a unique personal and existential significance; as a consequence of this relation with biblical revelation, together with the desire to account for the gospel in an easily comprehensible way, these truths are systematised into logical topics and expressed in our own words.

Because dogmatics is something which has been studied in the church for centuries on the basis of certain confessions, any new dogmatic reflection always originates in consideration of that which has already been written about the subject.

If we examine Holy Scripture, we notice between 300 to 400 references to angels.[6] Angels form an indispensable part of the history and message of salvation. Angels are directly connected to the first human couple, the history of the Old Testament patriarchs, the giving of the law, protection against danger, guidance, redemption as well as punishment meted out to the people of Israel. Angels are also associated with the actions of judges, priests, kings and prophets, the coming of Christ to this earth together with his ministry, resurrection and ascension. Furthermore angels are also mentioned in the life and ministry of the apostles and the return of Christ. This shows that the topic of angels is interwoven with other dogmatic topics. Therefore angelology can never be omitted from dogmatics. On the other hand, some dogmaticians persist that the topic of angels is something which originates in the mythological world and which has no place in dogmatics.[7]

It is indeed a pity that the existence and function of angels has been placed under suspicion by various erroneous representations. We think here

5 Regarding this new interest in supernatural phenomena, we refer to what Billy Graham has written in his book, *Angels: God's secret agents*, 1976, pp 13–26.

6 In the foreword of his book, *Angels in Heaven and Earth*, Stephen D Swihart writes, *inter alia*, the following: "References to God's angels appear over 375 times in the Bible ... This abundance of revelation on the subject demands recognition. There can be no full or accurate understanding of God's dealings with people if we should for some reason bypass or treat lightly this topic of Scripture."

7 P Tillich maintains that angels and demons are but mythological names for constructive and destructive powers which are ambiguously interwoven. They fight each other in one and the same person, but also in the same social community and in the same historical situation. *Systematic Theology II*, pp 41–50.

in particular of the heathen and mythological descriptions, the speculations of and wild claims made by certain theologians,[8] the unrealistic claims of people who are supposed to have had something to do with angels and the like.

These same things can, however, be seen as the obscuration and distortion of what is really true – something which is just a natural consequence of the fall of man. (The false is often the distortion of truth and can as such be an indication of something genuine which does in fact exist).

Because of the things mentioned above, it is essential to investigate what the Bible teaches us about angels.

2 WHAT IS AN ANGEL?

As a result of the question about what an angel is, different things come to one's attention. Foremostly let us consider the origin of the word.

2.1 The word *angel* and other appellations

The Hebrew word for angel, namely *malak* means messenger or agent (Gen 16:7; 1 Sam 29:9; Zech 1:9). It has the same source as the Babylonian word. (Biblical critics see cause in this similarity to suggest that the Hebrews got the concept *angel* from the Babylonians during the exile.[9] This view is also connected with their allegation that the Pentateuch only originated during that later period).

Sanskrit has the word *angiras* meaning a divine spirit. The Persian word *angaros* means messenger or courier. In Arabic we have the same word as in Hebrew, namely *malak*.

The Greek words as found in the New Testament are *angelos* (messenger) from which our word *angel* is derived.

Subsequently we have the Hebrew word *elohim* (Ps 96:4, 5) and *shadim* (Deut 32:17) which is translated as gods and devils or demons (Greek: *daimon*) respectively.[10] In Psalm 8:5 *elohim* is rendered *angels* in English (Ps 8:5).

In the New Testament other expressions are also used in connection with angels, such as authorities, powers, forces, thrones, principalities and spirits (Rom 8:38; Eph 6:12; Col 1:16).

The *morning stars* (Job 38:7) which rejoiced together at the creation of the earth, apparently also refer to angels. This also applies to the expression *sons of god* as mentioned in Job 1:6 and 38:7.

Psalm 68:17, 2 Kings 2:11 and 6:17 mention *chariots of God* or *chariots of fire* which can also be seen as references to angels or at least descriptions of their actions.

What appears from all these references to angels (holy as well as evil) is that they are supernatural beings who must be seen in relation to God himself.

8 Alan Richardson remarks: "In medieval theology, through the combination of biblical literalism with Greek philosophical speculation (Aristotle held that planets were superhuman living beings), angelology, the science of angels, was greatly elaborated; the various ranks of angels were held to constitute an essential link in the Great Chain of Being." *A Dictionary of Christian Theology*, 1969, p 8.

9 Cf Gustav Davidson, op cit, p 20.

10 Cf Grimm, *Greek-English Lexicon of the NT*, p 124.

2.2 Angels as spirits

Hebrews 1:14 says in connection with angels: "Are they not all ministering spirits sent forth to minister for those who will inherit salvation?"

The fact that angels are *spirits* means different things.
First and foremost angels do not have a material-physical body like humans or any other living creatures. Neither do they have a glorified body the same as that of the Lord Jesus at his resurrection (Luke 24:39) nor do they have the spiritual body which the believer will have after he has risen from the dead (1 Cor 15:44–50). This does not mean, however, that angels do not have a body at all or a particular form,appearance and personality. An angel is not just part of one or other impersonal ethereal matter. The portrayal of angels given us by the Bible is similar to people or other beings which can be depicted except that they have supernatural abilities.

Since angels are spirits, they cannot be observed by natural means or studied in a scientific-controllable way. They are not objects or phenomena from this world, but they are visitors from a domain which man himself cannot penetrate. We can only speak of angels because the Bible mentions them and because they act on behalf of believers. This prohibits us from speculating about angels or comparing them with anything that we know.[11] What we say about angels must either be directly taken from the Bible or must be deduced from what the Bible has said about them. Any knowledge which man might obtain as a result of angels' appearances or experiences which point to their actions must always be in agreement with what is found in Holy Scripture.

For the precise reason that angels are spirits and do not form part of our visible world, they are not subject to corruption and natural death as we know it (Luke 20:36). In addition, this implies that their survival is not assured by procreation. The Bible explicitly says that the angels do not marry (Matt 22:30). Therefore we have to accept that all angels were created in the beginning and that their number remains constant.

From what we read about angels in the Bible, we can most certainly deduce that neurological, physiological and other physical processes such as man possesses, do not form part of the existence of these spiritual beings. They did in fact eat and drink that which people set before them (Gen 19:1–3), but these substances did not become part of themselves. They are not dependent on the oxygen in the earth's atmosphere or on the food and drink procured here in order to survive. The cases where angels did eat should rather be seen as part of the way in which they made themselves known to people. (This is beside any liturgical significance that this might have had.) When Psalm 78:25 calls the manna provided for Israel in the desert, *bread of angels*, it does not imply that angels were dependent on this food for their survival. We should rather understand the expression *bread of angels* as a poetic expression which indicates that the manna was not produced by the Israelites, but that it fell from heaven.

Since angels are spirits and do not form part of our way of life here on earth, they are also not dependent on a physical dwelling place as we are.

11 During the time of the Reformation Calvin warned that theologians should not get involved in senseless speculations about angels. *Inst* 1, 14, 4.

However, the possibility is entertained by certain theologians that angels dwell on stars and planets in celestial space.[12] When the Bible speaks about the heaven where the angels are found (Matt 24:26; 28:2; Mark 12:25; Luke 2:15) or the air (Eph 6:12) where the evil spirits find themselves, this should not be seen as a scientifically verifiable area in this physical universe. Heaven and air, as it is used in connection with angels and evil spirits, refer to the region which transcends this visible world. Regarding the holy angels, the heavens is the region where they appear before the presence of God, the Father (Matt 18:10; Job 1:6).

Because angels enter our world from this transcendent place, things like speed and distance do not matter. Angels appear at a determined time and place and then disappear just as quickly. Precisely because they are spirits, neither matter nor distance between material things limits them (Acts 12:7–11).

Although angels are spirits, they can appear as visible and recognisable beings in this world, for instance in human form. For this reason Abraham, Joshua and others did not immediately realize that certain people with whom they were dealing, were not ordinary people, but were in fact angels. Hebrews 13:2 reads: "Do not forget to entertain strangers, for by so doing some have unwittingly entertained angels."

The form and manner in which angels appeared to people in biblical times always concurred with those peoples' contemporary way of life. In particular, we refer here to the clothes worn by angels, their weapons and the language spoken by them (Jos 5:13–15).

As spirits, angels exist on a superior level to man and they can do things which man would find totally impossible. Angels can be regarded as intermediaries between Christ (and thus also God, the Father) and man. In regard to man, we read in Psalm 8:5 "For You have made him a little lower than the angels, and You have crowned him with glory and honour." In connection with Christ's emptying of himself and suffering in the flesh, we read in Hebrews 2:9, "But we see Jesus who was made a little lower than the angels, for the suffering of death crowned with glory and honour, that He, by the grace of God, might taste death for everyone." In other words, the Lord Jesus was exalted above the angels and then for a short while placed on a lower level than the angels. The fact that an angel occupies a position between Christ and man, does not, however, imply that he is mediator between God and man. Christ alone is the Mediator.

2.3 Angels as creatures and servants of God

The angels are not gods beside God, but they are created to be servants and messengers of the Most High.

The time as well as the way in which angels were created are not mentioned in the Bible. There are, however, certain passages of Scripture which may be linked to their creation. In Genesis 2:1 we read: "Thus the heavens and earth, and all the host of them, were finished." The expression *all the*

12 Because the Bible refers to the sky and the stars also as heaven (Deut 28:26; Gen 1:14–17; Matt 24:29; Eph 6:12), A C Gaebelein maintains that stars and their planets are most probably the residence of angels. Op cit, p 11.

host of them (Hebrew: *tsebha'am*) can refer to angels in the context in which it is used in Genesis 2:1 (1 Kgs 22:19; Neh 9:6; Ps 103:20–21; 148:2). However it can also refer to stars (cf Deut 4:19; 17:3; 2 Kgs 17:16; Isa 34:14).

A more direct reference to the creation of angels is found in Colossians 1:16 where the following is said about Jesus Christ: "For by Him all things were created that are in heaven and that are on earth, visible and invisible, whether thrones or dominions or principalities or powers. All things were created through Him and for Him."[13]

As was already mentioned in Chapter 2, the creation narrative focuses on the preparation of the earth and man's appearance thereon. References to the stars as well as invisible beings such as angels, are only mentioned in the measure in which they fit into God's history with mankind.

Angels may not be worshipped because they are creatures and servants of God. When John wanted to worship the angel on account of its glory, the angel said: "Do not worship me! I am a fellow servant of you and your brothers, the prophets, and of them who keep what is written in this book. Worship God." (Rev 22:8). In Colossians 2:18 Paul rebukes those who want to worship angels.

As servants of God angels are also called messengers. (Although there are a few instances in the New Testament where the Greek for angel [angelos] indicates a human messenger (Luke 7:24; 9:52; Jas 2:25), as a rule it refers to angels. Angels are used to convey the presence, thoughts, will and actions of God regarding man. It is precisely this special and direct link which God has with angels that makes them very special beings. They are heavenly beings, distinct from earthly creatures. They represent a world which is above and different from this one (1 Tim 5:21; Heb 12:22). To be like an angel amounts to reflecting this transcendental world (Acts 6:15). When people are compared to angels, it implies the same as when comparing them with the divine (Gal 4:14). To be a spectacle to angels, implies being the same for all in heaven (1 Cor 4:9).[14]

3 THE FALL OF SOME ANGELS

Until now we have only spoken about angels in general, without referring to the distinction between the holy and fallen angels. In order to make a further discussion about angels meaningful, clear distinction should be drawn between these two kinds of angels. This brings us to the fall of Lucifer (the Satan or devil) and his angels.

The Bible does not mention how and when this fall of the angels took place. What we do find is that this fall is presented as fact.

The Bible mentions how the important angel named Lucifer (which means: the shining one or the morning star) together with a number of other angels rebelled against God. Together they were rejected by God and came under his judgement. However, they were not destroyed and for this reason since their rebellion they have raged against God and his creation (especially against mankind).

13 Gerhard Kittel says the following with reference to Col 1:16: "The angeloi can be reckoned with the *thronoi, kuriotetes, archai and exousias." Theol Dict of NT*, Vol 1, p 86.

14 Cf G Kittel, op cit, p 83.

In 2 Peter 2:4 we read about the angels who sinned and who consequently came under God's judgement. We also find this same idea in Jude 6 where the angels who did not keep their proper domain and who were therefore condemned by God are mentioned. (The context in which the Greek word *archei* – translated with *proper domain* – is used here, indicates: the original state, a state of worthiness or an area of authority.)[15]

God who through Jesus Christ created all things, visible as well as invisible, made them good and to his glory (Col 1:15–17). This same Christ had to come to this earth to destroy the works of one of his creatures who turned into a devil (1 John 3:8(b)). Thus there had to be an event, a fall, which took place whereby the angels changed into evil spirits.[16] Therefore the fall of the angels also explains how the first human couple were seduced to sin. Such deception is not possible unless it comes from outside the world of man, that is from the world of the angels (Gen 3:1–19).

The fact that the Bible tells us that a group of angels is warring against another group (Dan 10:13; Rev 12:7, 9) indicates that a serious problem must have arisen in the world of the angels. We read in John 12:31; Luke 10:18 and Revelation 12:7–9 how the prince of this world, namely Satan, was cast out of heaven. This clearly implies that he originally had to be in heaven when he forfeited that privileged status through his deeds.

Isaiah 14 and Ezekiel 28 gives us an allegorical presentation of the fall of Satan. (By allegorical the following is meant: The figurative way of expression whereby a person or object is described in terms of another person or object, better known by man and showing resemblance.) Thus in Isaiah 14 the parody intended for the king of Babylon, at the same time also presents a transcendental truth, namely the fall of Lucifer. The reference to these fallen angels (Satan), can especially be detected in the following words: "How you are fallen from heaven, O Lucifer, son of the morning! How you are cut down to the ground,You who weakened the nations! For you have said in your heart: I will ascend into heaven, I will exalt my throne above the stars of God; I will also sit on the mount of the congregation on the farthest sides of the north; I will ascend above the heights of the clouds, I will be like the Most High. Yet you shall be brought down to Sheol, to the lowest depths of the pit." (Isa 14:12–15).

The same also applies to what we read in Ezekiel 28. In the lamentation about the prince of Tyre, it is said, among others, that he was a cherub who was crowned with all kinds of glory. He was in the garden of God and was appointed on his holy mountain. He was like a god who had been created perfectly at the beginning. After this pride and violent rebellion entered his heart. This corrupted him and he was driven from the mountain of God and

15 Cf. *Grimm, Greek-English Lexicon* of NT, p 77, and also Louw and Nida, Greek-English Lexicon, Vol 1, p 478.

16 Karl Barth does not believe that the devil and his angels lapsed into sin, or that the evil angels were originally part of the holy angels. He maintains that the devil came from the domain of das *Nichtigen* (the Nothingness), that is the nothing which is contrasted against creation. *Kirchliche Dogmatik III*, 3, pp 608–623. Such a metaphysical view does not fit into the biblically clear and direct statements at all. The Bible states explicitly that the evil spirits were angels before their fall. We further notice that Barth has made of the so called *nothingness*, something so real that it has become a power that is in opposition to God and his creation. How is it possible that something which exists and is a power, at the same time be nothingness?

cast down from his previous glory. In all of this we see a description of Lucifer and his fall. (This turn of speech is something unique to Old Testament prophecy. In Psalm 2:7–9; 22:1, 8, 16; 110 we also find experiences of people being described as those which will apply to Christ one day. Here we encounter the so-called *Mehrgedanke*, that is, where the inspired author says more than what he knew or meant.)

The Bible does not tell us how large a number of the angels were involved in the fall. However, there are those who maintain that it was a third. Firstly they point out that stars sometimes refer to angels. Thereafter they quote from Revelation 12:4 where it is said about the fiery red dragon: "His tail drew a third of the stars of heaven and threw them to the earth."[17]

(In the following chapter we will deal more extensively with doctrine of fallen angels.)

4 THE MANNER IN WHICH ANGELS APPEARED TO PEOPLE

Holy angels serve as a link between God in heaven and man on the earth. As heavenly beings they also have to appear on earth in an appropriate form so that they can make contact with people.

4.1 Appearances in power and glory

The appearance of angels reveals to us something of their heavenly character. Although they usually appear as beings similar to people, their supernatural nature shines through strongly. Concerning their appearance we read that they appear in shining clothes (Luke 24:4; Acts 10:30) and also in snow-white garments (Matt 28:3). Sometimes the form of an angel can be terrifying to a human being, such as lightning (Judg 13:6; Matt 28:3). Sometimes angelic appearances go hand in hand with powerful natural phenomena. Thus it is mentioned that there was an earthquake (Matt 28:2), a blinding bright light (Acts 12:7), fire, dark clouds, lightning bolts and claps of thunder (Exod 19:16–18; Deut 33:2).

The effect of such angelic appearances is that the person experiencing them becomes terrified and helplessly falls to the ground (Dan 8:16–18; 10:16–17; Luke 24:4–5). Such a person feels as if he has come face to face with God himself (Gen 32:24–32). For this reason it was also necessary that such a person first had to be fortified by the heavenly being in order to receive the Lord's message.

4.2 The anthropomorphic character of angelic appearances

It can also happen that angels disguise themselves so well in ordinary human appearance that they cannot be identified immediately as heavenly beings. For those who encounter them, they appear to be like men. (It is striking that the masculine gender is always used when angels are mentioned, although they do not exist as male and female.) It is only after a while that people realise that they are dealing with heavenly beings (Gen 18:1–8; 19:1–5; Josh 5:13–15; Acts 12:7–11). For this reason the writer to the

17 Cf B Graham, op cit, pp 65, 66.

Hebrews says that people have received angels unawares (Heb 13:1).

The anthropomorphic character of angelic appearances is also clear in that they speak a language comprehensible to the people of that time. As has been already mentioned, their clothing, weapons in their hands, musical instruments played by them, the food they eat et cetera were all familiar to the people of the times. Consider Joshua 5:13; 1 Kings 19:5, 6 as well as John's visions of angels on the island of Patmos.

5 KINDS OF ANGELS

According to the Bible the holy angels are not all of the same kind or rank. This diversity in the kingdom of the angels also reveals God's eternal wisdom and plan – something of which only a small part has been revealed to us.

5.1 The Angel of the Lord

Firstly we give attention to the Angel of the Lord before turning to a discussion of the different kinds of angels.

When we give attention to the way in which the Bible speaks of the Angel of the Lord; how he implies the presence of God, yes, this Angel is the form assumed by the Almighty when he makes himself known, then we cannot but see this as God's revelation through Christ. We consider some examples in this regard.

In Genesis 22:11 the Angel of the Lord addresses Abraham. In verse 12 the same Angel says: "... for now I know that you fear God, since you have not withheld your son, your only son, from Me." This *me* refers to God and to the One who commands Abraham to sacrifice his son in verses 1 and 2. At the same time the pronoun me takes the genitive form of the Angel of the Lord who appeared to this patriarch.

In Exodus 3:2 we learn how the Angel of the Lord appeared to Moses in a burning bush. Verse 4 tells us that it is the Lord God who addresses him from the burning bush. The sixth verse expressly states who the *I* is who is addressing Moses. We learn this from the following words: "I am the God of your father – the God of Abraham, the God of Isaac, and the God of Jacob." From verse seven until chapter 4:17 we find further confirmation that the Angel of the Lord who spoke with Moses from the burning bush was actually the Lord God himself. Genesis 16:7–12 declares that the Angel of the Lord found Hagar in the wilderness and addressed her. Here the angel said to her: "I will multiply your descendants exceedingly, so that they shall not be counted for multitude" (verse 10). This *I* must be the same God who uttered the same words about the same matter in Genesis 17:20. We then also see how Hagar herself witnesses that it is the Lord God who spoke to her (Gen 16:13).

At a later stage another meeting took place between the Lord and Hagar (Gen 21:17–19). Also in this instance the Angel of the Lord who spoke to her was the Lord God himself (cf verse 18).

Genesis 31:11 mentions that the Angel of God appeared to Jacob, and in verse 13 the same One says: "I am the God of Bethel, where you anointed the pillar and where you made a vow to Me."

This unity between the Angel of the Lord and the Lord God himself is also apparent from other passages of Scripture, for instance Judges 2:1–23; 6:12–15; Psalm 34:8. We find a similar unity between the Angel of the Covenant and the Lord of Hosts. In Malachi 3:1 we read: " '... and the Lord, whom you seek, will suddenly come to His temple, even the messenger of the covenant, in whom you delight. Behold He is coming,' says the Lord of the hosts."

However, on the other hand, it appears we should distinguish between the Lord God and the Angel of the Lord. We notice this from references such as Judges 5:23. This is the case where the angel of the Lord refers to the Lord. We also refer to Judges 13:9; where a distinction is made between God who listened and the Angel of God who was sent. Also see 2 Samuel 24:16; this is a case where the Lord addressed the angel who brought about the destruction. That angel was apparently the same one who was known as the Angel of the Lord at the threshing floor of Arauna (1 Chr 21:16–18). In these passages we see that the Angel of the Lord is an indication of the Lord and God. Most probably the Being described in Ezekiel 8:2 is also the Angel of the Lord, especially as we notice that it was the hand of the Lord which fell on the prophet in the first instance.

We find another similar description with regard to the Lord, the Angel of the Lord and angels in Zechariah 1:8–17.

A possible pattern can be detected in the passages about the Angel of the Lord. When a matter concerns the Lord as such, the terms *Lord* or *God* are used. When the form in which the Lord appears to man or which he uses to reveal himself to man is mentioned, the collocation *The Angel of the Lord* is used. Thus we read in Genesis 21:17–19 that God heard the voice of the boy, but it was the Angel of the Lord who called Hagar, and once again it was God who opened her eyes.[18]

Our conclusion is that the unity between the Angel of the Lord and the Lord God himself, as well as the distinction between the two, is a reference to the relationship between Jesus Christ and God, the Father. On the one hand Christ is one with the Father as the triune God. On the other hand Christ is distinct from the Father, in the sense that He is the Son.

The mention of *an* (notice: not *the*) angel of the Lord (God) as found in the New Testament cannot be equated with Christ. Notice the following examples: Matthew 1:20; 2:13, 19; 28:2; Acts 5:19; 8:26; 12:7, 23; Revelation 1:1, and so on. We get the impression that in the Old Testament, where Christ has not yet appeared as the Messiah and the Saviour, He is referred to as the Lord. In the New Testament, after Christ has appeared as the Messiah and Saviour, he is no longer referred to as the Angel of God. The occasion when the glorified Lord Jesus reveals himself to Saul (later Paul) from heaven, He declares that He is Jesus, the Nazarene (Acts 22:8).

5.2 The archangel

We read about Michael, the Archangel, in Jude 9. The prefix *arch* is derived from the Greek *archi*, which when linked as a prefix to a word renders the meaning: head, most important, first class or rank, such as archbishop, archenemy, archangel, and so on. In combination with angel, archangel thus

18 The same notion is expressed by G von Rad. *Theol Dic NT*, 1, p 77.

means a chief of the angels or a very important angel. The Hebrew word which renders the same concept as archangel is *sar* such as found in Daniel 10:13, 20; 12:1. There are also other translations for this Hebrew word, for instance *prince*. Irrespective of which translation one may choose, sar indicates an angelic being as it appears in the passages mentioned.

The fact that Michael is specifically called an archangel (Jude 9) or one of the most important princes (Dan 10:13; 12:1) does not necessarily imply that he is the only one. The mere fact that he is called one of the chief princes (Dan 10:13) supposes that there must be others who fall into the same class. The same Hebrew word used to refer to Michael is also used to refer to the princes of Persia and Greece (Dan 10:20–21).

In connection with the second coming we read that Christ "... will descend from heaven with a shout, with the voice of an archangel, and with the trumpet of God." (1 Thess 4:16). This is not a reference to the deeds of a specific archangel (e.g. Michael), but to the loud and authoritative summons or command of an archangel. From the above arguments, it is very probable that there is more than one archangel. Thus the angel Gabriel could possibly also be considered an archangel.

Extra-biblical sources maintain that there are seven archangels and that each commands many thousands of ministering angels.[19]

The mention of the archangels suggests that the angels stand in a particular order of rank. There are those who lead and those who follow.

5.3 Seraphs

The word *seraphs* (Hebrew: *seraphim*) literally means: the *burning ones*, derived from the verb saraph which means: to burn or to consume.

The only place in the Bible where we read about seraphs is when the prophet, Isaiah, saw the vision of the Lord on his throne. "Above it stood seraphim; each one had six wings: with two he covered his face, with two he covered his feet, and with two he flew. And one cried to another and said: Holy, holy, holy is the Lord of hosts; The whole earth is full of His glory!' " (Isa 6:2–3).

When Isaiah experienced all this, he called out: "Woe is me, for I am undone! Because I am a man of unclean lips, And I dwell in the midst of a people of unclean lips; For my eyes have seen the King, The Lord of hosts."

As Isaiah realized his own uncleanness and unworthiness, a seraph approached him with a coal in his hand which he had taken from the altar with a pair of tongs. He touched Isaiah's mouth with it and said: "Behold this has touched your lips; Your iniquity is taken away, and your sin purged."

Two things emerge forcefully from this vision: God's holiness and the fire which consumes all impurity.Not only do the seraphs declare God's holiness in his presence, but they also cover their own faces and feet in holy reverence. God's holiness is like a burning and consuming fire in the presence of which no sinful and imperfect creature can exist. Subsequently, it is this truth which must be linked to the significance of the seraphs: the burning and consuming holiness of God.

19 In the apocryphal book of Enoch one finds the following names of the seven archangels: Uriël, Raphael, Raquel, Micael, Zeragiél, Gabriél and Remiél. Gregory (theologian from the 4th century AD) presented the following names: Migael, Gabriel, Raphael, Uriél, Simiél, Orifiél and Zegariél. Cf Gustav Davidson, op cit, pp 338, 339.

There are theologians who do not include the cherubs or the seraphs among the angels. The reason for this exclusion is because they do not have the appearance of human beings as is the case with the other angels.[20]

Besides the wings a seraph apparently still has a form which can be associated with the human form. He also has a face as well as hands and feet. But why should such a being have to assume a human form to qualify as an angel?

5.4 Cherubs

Cherubs, like the seraphs, are a particular class of angels. Angels, in the ordinary sense, are the beings whom God uses to convey messages to people and to do certain deeds in this world. Thus they usually appear in human form. As opposed to this, cherubs (just like seraphs) form part of the visions of the Lord and his holy throne revealed to those whom he called to a special ministry. Here we think for example of Ezekiel (Ezek 1:9; 3; 10:1–16). Depictions of cherubs also appeared on the entrances to the temple of Solomon, as well as on objects, doors and walls of this sanctuary (1 Kgs 6:35; 7:29; Ezek 41:18, 20, 25).

We are not certain of the origin and the meaning of the word *cherub*.[21] Portrayals suggesting cherubs also appeared in ancient Babylon, Assyria and Egypt. In all these portrayals it appears that the cherubs were the beings closest to the deity who served as guards at the sanctuary to protect the gods. There is a possible reference to cherubs who guarded the entrance to the Garden of Eden – with the Tree of Life (Gen 3:24).

Two things become very clear from what we read in the Bible in connection with the cherubs. Firstly they reveal the majesty and glory of God. Secondly they serve as guardians of that which is holy.

5.4.1 Cherubs as revelations of God's majesty and glory

The visions in which the cherubs appear reveal something of the majesty and the glory of God's presence. For this reason Ezekiel describes the visions which he had of the beings or cherubs as *visions of God* (Ezek 1:1ff). Hence we read of the name of the Lord of Hosts who was enthroned above the cherubs (2 Sam 6:2; 2 Kgs 19:15; Ps 80:2) and of the glory of the God of Israel which ascends from the cherub (Ezek 9:3; 10:19; 11:22).

Although the forms of the *beings* or the cherubs are human (Ezek 1:5), they are not limited to that form. (We should note that the *being* whom Ezekiel saw at the Chebar River is the same as that which is later called a cherub.) (Ezek 10:15.) Precisely because God's majesty and glory is concerned here, the description of the cherub in this instance transcends

20 G von Rad remarks about angels: "Externally these heavenly messengers are depicted by the Hebrews in human form and without wings. In OT literature there are also references to beings in the form of animals, the seraphim and cherubim, though these are not to be counted as malakim." Op cit, p 80.

21 Some maintain that *cherub* (Hebrew: *kerub*) is derived from the Babylonian word *karuabu*, meaning great, mighty and awesome. Others again propose that it is derived from another Babylonian word, namely *kaaribu*, meaning: intercessor, worshipper, one paying homage. Cf Prof Dr F W Grosheide, *Bijbelse Encyclopedia*, 1960, pp 101, 102.

human form. We read that every cherub has two pairs of wings. With the one he covers his body and with the other pair, he flies (Ezek 1:6, 11, 23–24). We also read: "Their legs were straight, and the soles of their feet were like the soles of calves' feet. They sparkled like the colour of burnished bronze" (verse 7). In addition everyone has four faces, namely that of a human, a lion, an ox and an eagle (Ezek 1:10). The cherubs are subsequently described as follows: "... their appearance was like burning coals of fire, like the appearance of torches going back and forth among the living creatures. The fire was bright, and out of the fire went lightning. And the living creatures ran back and forth, in appearance like a flash of lightning" (verses 13 and 14). Subsequently verses 15–18 state that with them were turning wheels of which the rims were full of eyes. Revelation 4:5–8 describes beings which resemble the cherubs, except that they, just like the seraphs, had six wings and called incessantly: "Holy, holy, holy is the Lord God." There is also a similarity between the cherubs seen by Ezekiel and the angel who appeared to Daniel (Dan 10:4–6).

We must accept that the different things said about the cherubs have distinct liturgical significance. They symbolise something of God's majesty and glory.[22]

The total difference in nature and even the strangeness of these revelations of God through the cherubs in fact serve to stress the mystery and the splendour surrounding him. This does not concern everyday matters which people observe or imagine with regard to God. After all, this is about the supernatural, something about God himself.

5.4.2 The cherubs as guardians of that which is holy

The first mention of angels in the Bible is found in Genesis 3:24. Their task was to prevent anybody from having access to the Garden of Eden and the Tree of Life. When Adam and Eve were still without sin, they could live in the garden (Paradise) where they could enjoy fellowship with God. After they had become unclean and corrupt through sin, they could no longer enter into this holy place.

The cherubs' task of protecting and guarding is also portrayed with regard to the ark of the covenant. Above the mercy seat under which the ark was kept were the images of two large cherubs. They spread their wings protectively over the ark – the place where God revealed himself in his glory and holiness (Exod 25:19, 20, 26; 1 Kgs 6:24–35; 2 Chr 3:14). In 1 Kings 8:7 we read: "For the cherubim spread their two wings over the place of the ark, and the cherubim overshadowed the ark and its poles." The writer to the Hebrews speaks of "... the cherubs of the glory which overshadowed the mercy seat." (Heb 9:5.)

Only the high priest could enter this holy of holies once a year. He could

22 J Barton Payne remarks regarding the cherubs: "The appearance that they assumed for conveying God's revelation was human (Ezek 1:15), not that of the mythological winged sphinx or Assyria-type lion, as is claimed by liberal historicism. Ezekiel 10:20, 21 mentions wings and four faces, signifying the wisdom and power of the cherubim – they are able to see everything that goes on, in all four directions at once! ... God made their visible appearances to differ, as each occasion might best be served." *The Theology of the Older Testament*, 1962, p 286.

only do this after he had purified himself, entering with the blood of a sacrificial animal. He had to sprinkle this blood on the holy objects as an atonement for the sins of the people, as well as for himself.

The guarding and protective function of the cherubs emphasizes the holiness of God's heavenly throne. Nothing unclean may appear there. We read in connection with Lucifer: "You were the anointed cherub who covers; I established you; You were on the holy mountain of God; You walked back and forth in the midst of fiery stones ..." However, after he sinned he was driven from this holy place (Ezek 28:14–16).

5.5 Ordinary angels

Besides the seraphs and the cherubs, there are also beings which are only known as angels.

With regard to this innumerable host of angels, the emphasis falls on what they are doing. This includes: conveying God's messages to man, protection and rescue of certain people as well as the people of Israel, events in connection with the coming of Christ and his ministry on earth, care and encouragement of those in the Lord's service, battle against evil angels, the announcement and the execution of the judgements of God, the return of Christ, and so on.

Many of the deeds of the angels are mentioned in the visions of John recorded for us in the Book of Revelation. In this book we also find a few instances in which the names of angels are singled out. Hence we read in Revelation 18:10 about a star or angel called Wormwood. He is called this on account of the bitterness which he caused people. We read further in Revelation 9:1 about another angel to whom was given the key to the bottomless pit. When he unlocked this bottomless pit, creatures similar to locusts emerged. These beings who tortured people were under the command of the *angel of the abyss*. His Hebrew name was Abaddon and in Greek he had the name of Apollyon (Rev 9:11).

It is a fact that the Bible links many situations and events to angels. However, it gives us no right to relegate everything which happens to a person, whether good or evil, to the action of the angels. Nevertheless this is what happened in the Middle Ages. There were those in the church who proclaimed that there are hundreds of kinds of angels. Every angel was considered responsible for specific events on the earth. Consequently there was the angel of the harvest, an angel of fortune, an angel who brought death, an angel who caused pestilence, an angel (or devil) for every kind of illness, angels for certain times of the year and of the day, and so on.[23] Not only are such theories pure speculation, they are nothing less than superstition. Such teaching does not belong in the church.

All that we know is that God is in fact involved with this world and mankind. According to his own purpose He can use angels to execute his will on earth. Whether God has separated certain angels for certain tasks and which angel has been allocated a specific task are all things which the Bible does not reveal to us.

23 Cf Gustav Davidson, op cit, pp 339–350.

6 ANGELS IN THE SERVICE OF GOD

Hebrews 1:14 says the following in connection with angels: "Are they not all ministering spirits sent forth to minister for those who will inherit salvation?"

We have to accept that angels were created in the first place to act to the honour and glory of God. In the scripture passage mentioned above, our attention is only focused on the human perspective. It tells us that the angels are there to act on behalf of the Lord for the sake of man.

We firstly pay attention to the service which angels render God on his behalf. After this we shall focus more specifically on the relationship between Jesus Christ and the angels, as well as between the Holy Spirit and the angels.

6.1 The involvement of angels in God's revelations to man

In God's revelations to man (theophanies), a distinction is not always made between the Lord and his angels. This distinction is sometimes vague because the holy angels do not convey their own but God's presence, will and his deeds to people. In any case when we deal with the holy angels we are concerned with God.

We can accept that whenever the Bible speaks of a direct divine intervention in this world, angels are most probably involved. Although angelic involvement is not always mentioned in a particular situation, it may be determined from later revelations. We refer to a few examples.

With regard to the judgements which came upon the Pharaoh of the Egyptians for the sake of the Israelites, no mention about angels is made (Exod 7–11). In Psalm 78:49 we find a further revelation in connection with these judgements: "He cast on them the fierceness of His anger, wrath, indignation, and trouble, by sending angels of destruction among them."

In the narrative about the giving of the law on Mount Sinai (Exod 20), nothing is mentioned about angels. In Acts 7:53 Stephen declares that the Israelites "... have received the law by the direction of angels ..."

The deeds of angels are never intended to draw man's attention to them but to God himself. Thus it is not surprising that an angel may sometimes appear to people who only later realize that it was in fact an intervention of a heavenly being (cf Acts 12:7–11; Heb 13:2).

It is only when the Bible slightly lifts the veil from the invisible world, that we can see what a major role angels play in this world.

6.2 The characteristics and the abilities of angels

What the Bible says about the deeds of angels, presupposes that they must also possess supernatural characteristics and abilities.

Since angels are spirits they enjoy a more direct relationship with God than man who is flesh. Angels do not have any material and physical characteristics such as is the case with man and which create a distance between him and his Creator. The angels are also not limited as man is on account of his corporeality and materiality. For this reason there is some excuse for man's wrongdoing, or better said, grounds to plead for God's grace and mercy. The Bible implies something like this when it refers to the *weakness of the flesh* (and mortality), together with the fact that man is but dust (Ps 78:38–39; 103:14; Matt 26:41; Rom 6:19; 7:14; 1 Cor 3:3; 2 Cor 4:11).

In the case of angels this is quite different. The distance between God and us and the weakness we experience do not apply to the angels. This means that an angel either obeys God in his totality or is against God in his totality. It can also be expressed differently: In the case of the holy angels the wisdom and the power and the splendour of God can be revealed through them explicitly and untrammeled. In contrast the fallen angels are totally rejected by God. Even though He continues to sustain them, the fallen angels are, however, an anti-god or negative force. They are destined for eternal judgement, although they are still tolerated. The Lord does this because He controls the evil force in such a way that his goodness, grace and glory can be confirmed thereby. This happens in that He saves man from the power of evil. In his battle with and victory over evil man gets to know God more and more and thus become conformed to Him. It is for this reason that the destruction of the evil angels do not immediately take effect.

On the grounds of what we have said regarding the angels as spirits, we cannot actually talk about the characteristics of holy angels as we do when referring to people. In contrast with mankind, as has been mentioned already, angels are not described in terms of the physical and corporeal nor are they thus limited. Moreover, angels do not have a specific dwelling place somewhere in the (to us) visible universe, such as is the case with humans. In addition they do not comprise a generation where knowledge, insight and tradition are passed over from one generation to another, from parent to child. In brief, the external and the inherited factors which work to determine the human personality, do not apply to angels.

Angels are spirits operating from heaven where they are directly under God's control. Because they are servants, agents and messengers under God's immediate control, they do not employ any abilities apart from God's direct working through them. Everything that angels do are revelations of God himself.

The wisdom, knowledge, authority, power et cetera which angels reveal (Matt 18:10; 2 Sam 14:20; Ps 103:20; 2 Thess 1:7; Rev 3:5; 14:10) do not belong to them; they are only creatures totally obedient to God. What a holy angel says or does actually amounts to God himself saying or doing that.

The abilities of angels are thus determined by their ministries, that is, in accordance with the mandate they have received from God. On their own they are mere creatures with all the limitations associated with this fact. In the execution of their tasks the unlimited and the divine, however, comes into operation.

This also applies to the multitudinousness of the angels (Matt 26:53; Heb 12:22; Rev 5:11). The vastness of the angelic army is presented to us humans not to be impressed by angels as such, but that we may become aware of God's majesty and power.

When we speak thus about angels, we do not mean that they are but instruments without a will in the hand of God. As has already been mentioned, the Lord created them – as in the case of man – with an own will and commensurate responsibility. They are, in other words, morally-conscious beings who are accountable to their Maker. God granted them room to remain faithful to Him or even to turn against Him. The fall of the angels testifies to this.

Furthermore, this implies that the holy and chosen angels – those who are faithful to God – do not serve their Creator because of external coercion, but on the basis of an inward desire. This makes the ministry of angels so particularly wonderful and beautiful.

The accountability of angels should not be understood in the sense that God has been abandoned to the will of the angels. When the angels serve God, they are only doing that for which they have been created. The reason why some angels departed from this original principle for which they were created and turned against their Maker is something which is totally incomprehensible. (The illogical and incomprehensible nature of sin has already been discussed in the chapters on the harmartology in Part II.)

6.3 The worship and reverence of God

As with any other creatures, angels have been created with a view to God's glory. The Lord God shares his glory with the angels, who, in turn, glorify Him. This worship and glorification of God is clearly evidenced in the case of seraphs and cherubs, as we have already seen.

In Psalm 103:20 and 21 David calls on all but in particular, the angels, to worship God: "Bless the Lord, you His angels, who excel in strength, who do His word, heeding the voice of His word. Bless the Lord, all you his hosts, you ministers of His, who do His pleasure." The same summons is also found in Psalm 148:2.

The vision of the four living beings (possibly seraphs or cherubs) before the throne of God shows how they incessantly call: "Holy, holy, holy, Lord God Almighty, Who was and is and is to come!" (Rev 4:8).

In Revelation 7:11–12 John sees how angels, together with the martyrs around God's throne, worship and glorify him.

To be able to be in direct contact with God and to glorify him with the glory which he shares with us points to the highest ministry imaginable. The salvation and glorification of people is precisely for the purpose of worshipping and glorifying God together with the holy angels (Eph 1:5–6; 2 Cor 3:18; 4:15–17; 2 Thess 2:14; 1 Pet 1:6–7; 5:1, 10; Rev 14:1–3).

6.4 The service of angels before the throne of God

Beside the worship and reverence of God, angels also have a liturgical function before his throne. Because this is mentioned in visions, they obviously comprise symbolic descriptions. However the fact is that this is nevertheless a reference to a certain ministry rendered by the angels in heaven.

In the same way in which the priests of old executed certain tasks in the temple, so also do angels execute certain tasks in the heavenly sanctuary (Isa 6:5–9).

In Revelation 8:3–4 an angel lays the prayers of all the saints in a golden censer mixed with incense before the throne of God. We also see how an angel gives commands to John and thereby conveys certain truths to him (Rev 10). Furthermore we read how an angel shows John the new Jerusalem and thereafter measures the city, its gates and wall with a golden measuring stick (Rev 21:1–17).

Some take the following as proof that angels sit in the same council as God: Psalm 89:5, where the Lord's faithfulness is extolled in the gathering of the saints, and 2 Chronicles 18:18, where the prophet, Micaiah says: "I saw the Lord sitting on His throne, and all the host of heaven standing on His right hand and His left".[24]

6.5 Angels as bearers and interpreters of God's message

We have already referred to Hebrews 1:14 which says that angels are serving spirits sent on behalf of those who would inherit salvation. Hebrews 1:7 says the following in connection with Psalm 104:4: "And of the angels He says: Who makes His angels spirits and His ministers a flame of fire.' "

Thus there are also numerous examples in the Bible of angels taking the Lord's message to people. We refer to some of these.

An angel appears to Gideon with the command to save the Israelites from the hands of the Midianites (Judg 6:12–22). Angels brought messages to Abraham in connection with the birth of Isaac and the destruction of Sodom and Gomorrah (Gen 18). We also read about an angel who appeared to Abraham and gave him instructions concerning his son (Gen 22:11–12).

Angels brought a message to Lot and his family to flee from Sodom and Gomorrah (Gen 19:1–17) We read how the Lord made certain promises to Jacob through angels in a vision in the night (Gen 28:11–22). We also read how an angel of the Lord commanded Jacob to move away from Laban (Gen 31:11–13). Acts 7:38 mentions an angel who conversed with Moses on Mount Sinai. We also read how the Law was given to the people of Israel by means of the intermediacy of angels (Acts 7:38, 53; Gal 3:19; Heb 2:2). In Daniel 8:16–26 we read how an angel appeared to this man of God and in a vision gave him a message about the future of certain nations. After this the same angel explained this same vision to Daniel. In the subsequent chapter the future of the nation of Israel was conveyed in the same way (Dan 9:20–27). In Acts 8:26 an angel of the Lord commanded Philip, the evangelist, to go and wait next to the road to Gaza, where he would meet the Ethiopian court official and lead him to Christ. God used an angel to convey the revelations to John on the island of Patmos (Rev 1:1; 22:6–16). The same John sees "... another angel flying in the midst of heaven, having the everlasting gospel to preach to those who dwell on the earth – to every nation, tribe, tongue, and people ..." (Rev 14:6). In the New Testament angels bring different messages about the coming of Christ to this earth – something to which we will return later.

6.6 Angels who protect and rescue people

The Lord himself ensures that his kingdom is revealed in this sinful and evil world and that it will ultimately triumph over everything. With regard to this

24 J Rodman Williams writes regarding Ps 89:5: " 'The holy ones' are unmistakably angels who form a council about God." Regarding 2 Chron 18:18 he remarks: "There was then a conversation between God and those around Him about a course of action to be taken. The right hand and the left here suggest some kind of organisation, but the most relevant matter is that the host of heaven, the angels, form a council in relation to Almighty God." *Renewal Theology*, 1988, 1, p 187.

He also determines that the call and ministry which He has for his servants, both men and women, will be performed on earth.

However, believers are continually subject to the onslaught of the evil one and they cannot resist him in their own strength. For this reason God often intervenes directly, among others through his angels in order to protect people and rescue them. We often read about this in the Bible.

In Psalm 91:10 and 11 the following promise is given to those who take refuge in the Lord – a promise which also refers to the coming Messiah: "No evil shall befall you, Nor shall any plague come near your dwelling: For He shall give His angels charge over you, To keep you in all your ways."

Jacob, the ancestor of the people of Israel, testified of the angel who delivered him from every adversity (Gen 48:16). The deliverance of Israel from Egypt took place through an angel sent by the Lord (Neh 20:15–16, cf also Ps 78:49). An angel of the Lord delivered Jerusalem from certain destruction when he destroyed 185 000 Assyrian soldiers in one night (2 Kgs 19:35). In 2 Kings 6:8–23 we read how the prophet Elisha and the people of Israel were saved from the Aramean armies when an angel army (horses and chariots of fire) appeared on the scene. This is a literal fulfilment of the promise in Psalm 34:7: "The angel of the Lord encamps all around those who fear Him, And delivers them." Daniel testifies how an angel of God protected him in the den of lions (Dan 6:23).

The enemies of the early church tried to either kill all the key figures amongst the believers or cast them into prison. In this way they would have tried to destroy the testimony concerning Christ. We read how an angel of the Lord on more than one occasion led the apostles from well guarded prisons (Acts 5:17–20; 12:1–10).

6.7 Angels who cared for and encouraged people

The ministry performed by angels on behalf of people in biblical times also includes temporal care and encouragement. We refer to a few examples.

Hagar and her son were sent into the wilderness where their water supply ran out. When it seemed that Hagar's son would die of thirst, the angel of God spoke to her. He showed her a well with water and above all assured her that her son's posterity would become a great nation (Gen 21:14–21). Elijah, discouraged by the threats made by the godless Jezebel, lay under a broom tree in the desert and longed for death. This same angel then supplied him with warm bread and water so that his body could once again be strengthened. Furthermore the angel told him to go to Mount Horeb. Strengthened by the food provided for him, he was able to walk for forty days and nights to his destination (1 Kgs 19:1–8).

While Paul was being sent as a prisoner by boat to Rome, they encountered a great storm. This persisted until the boat started sinking. At this stage an angel of God appeared to him and said: " 'Do not be afraid, Paul; you must be brought before Caesar; and indeed God has granted you all those who sail with you.' " (Acts 27:23–24)

6.8 The holy angels are conducting a struggle against evil angels

On account of the fall of the angels a war ensued in the spirit world. Revelation 12:7–9 puts it this way: "And war broke out in heaven: Michael and his angels fought with the dragon: and the dragon and his angels fought, but they did not prevail, nor was a place found for them in heaven any longer. So the great dragon was cast out, that serpent of old, called the Devil and Satan, who deceives the whole world; he was cast to the earth, and his angels were cast out with him."

Satan and his angels obtained control over this world through sin (1 John 5:19). The Lord, however, did not allow his purpose with man and mankind on earth to be disrupted thereby. Because of this, Christ came to this world to destroy the works of the devil and to reveal the kingdom of God on earth. Although the Evil One was cast out of heaven, he, however, still has a powerful influence on earth.

Evil spirits can, however, only act within the limits allowed by God. They cannot destroy the kingdom of God, as revealed in the believers of all times and, in particular, in the church of the last days. That which the Lord did through his holy angels comprises a way whereby this limitation is imposed on the Evil One.

The Bible indicates clearly this struggle between the holy and evil angels. We read in Joshua 5:13–15 about the *Commander of the Lord's army*. This shows that angels can also act as an army of soldiers.

In Daniel 10:13 we read how the angel of the kingdom of Persia opposed an angel of the Lord, so that the latter had to call on the archangel Michael for help. We also read about this struggle in verses 20 and 21: "Then he said Do you know why I have come to you? And now I must return to fight with the prince of Persia; and when I have gone forth, indeed the prince of Greece will come. But I will tell you what is noted in the Scripture of Truth. No one upholds me against these, except Michael your prince.' "

We read about the struggle between the archangel and the devil over the body of Moses in Jude 9.[25]

John's vision also refers to this struggle between the holy and the evil angels (cf Rev 20:1–3). In this passage we see how that an angel bound the snake who is the devil and Satan with an enormous chain. Thereafter he cast him into the abyss and sealed it.

The final outcome of this struggle will come about when Satan, together with all his angels, is cast into hell (Rev 19:11–21; 20:10).

6.9 Angels who exert an influence over the nations

We believe that the universe, the world, man and all that is on earth are not merely machines or robots which can function without the decree of a Higher Power. God created all things so that the visible things originated from the invisible (Heb 11:3). Furthermore, on the basis of Holy Scripture we believe that God sustains everything and that nothing takes place without his will or permission. God remains Lord over all. We can accept that in

25 In this regard Stephen Swihart writes: "Jude apparently gathered his data here from a non–canonical book entitled The Assumption of Moses'. In this account of Moses' funeral it is stated that Michael is commissioned to bury the body of the great patriarch. Satan on the other hand, opposed the action on two grounds: 1. He asserted supremacy in the matter and claimed that the body therefore, belonged to him. 2. He accused Moses of committing murder, making him one of his own disciples. The attempt on Satan's part, however, availed nothing. Still the episode provides us with another glimpse of how Satan wages conflicts in the sky." Op cit, p 93.

the execution of this control, invisible forces in the form of angels also play a distinctive role. Not only does this apply to individuals, but also to communities and nations.

There is good reason to believe that angels are involved in the rule and the fate of nations on earth. Although not much is mentioned about such influence of angels in this connection, the Bible does indicate this.

1 Kings 22:19–40 describes how an angel (spirit) was sent by God to certain prophets. The angel had to influence them in such a way that they would convey the wrong predictions to the godless king Ahab. On the strength of these predictions Ahab advanced against the enemy and was killed in the process. Humanly speaking, this was just another casualty which could be expected in the course of battle. In reality, however, this was the execution of the judgement of God over him.

Exodus 17:8–16 clearly shows us that it was the Lord himself who brought about Israel's victory over the Amalekites. According to external and human factors the outcome of the battle was merely determined by the soldiers, the weapons and the strategy used. Any connection between Moses' raised hands to God and the course of the battle is, humanly speaking, just not possible. However, in this battle the help of angels can be assumed, just as is expressly mentioned in the case of other battles (2 Kgs 6:15–23; 19:32–36). In the same way the Lord God used angels to punish the nation of Israel when they became unfaithful to him (2 Sam 24:11–17).

The fact that angels are also described as principalities, thrones, powers, forces and world rulers (Eph 1:20–21; 3:10; 6:12; Rom 8:38; Col 1:16; 2:15) is an indication that they act in a ruling capacity in this world.

We have already pointed out that Daniel 10:13 and 20 mention the king, prince or angel of Persia and Greece. This gives us reason to believe that there might be such angel–princes for other nations. Michael is described as the prince of the people of God in Daniel 10:21. This presupposes that angels have influence, in one way or another, over those nations, either as holy or as evil forces. We do not want to say that every nation has an angel who is appointed to care for its interests. The Bible gives no grounds for the pagan belief that each nation has national and political guardian angels.

Only in the hereafter shall we fully grasp (1 Cor 13:12) the importance of the role played by angels in the destiny of nations and the extent of their influence on nature and earthly events.

6.10 Angels and the end of time events

The Bible mentions considerable angelic activity in connection with the end of time events.

First and foremost we see how the Lord sent an angel to John on the island of Patmos to inform him (and via him, the entire church) more completely about the end of time events (Rev 1:11; 22:6). Subsequently we see what an important role angels played in the visions presented to this apostle.

Indications of angelic activities already occur at the end of an individual's earthly existence. We read about a deceased believer taken by angels to the Lord and the blessed in heaven. Whether this is a parable or represents reality, the Lord Jesus presents us with the following in Luke 16:22 : " 'So it was that the beggar died, and was carried by the angels to Abraham's bosom. The rich man also died and was buried.' " The way in which the prophet Elijah is raptured into heaven also suggests angelic activity (2 Kgs 2:9–12).

Angels are apparently involved in the gathering of the chosen at the end of time. The Lord Jesus describes it thus in Matthew 24:31: " 'And He will send His angels with a great sound of a trumpet, and they will gather together His elect from the four winds, from one end of heaven to the other.' " However, we may not deduce from this that the angels' task is to raise the bodies of deceased saints to life.[26]

The Bible foretells that tribulation will come upon the world with the express aim of summoning sinful humanity to repentance. In the visions of the apostle John, we see that angels have much to do with these plagues and the misery which will come upon the earth (Rev 6–19).

The Bible also teaches that angels are involved in carrying out God's judgements. We have already seen how in biblical times angels took part in carrying out God's judgements. We have seen something of this in the way in which the enemies of Israel were defeated. Furthermore, we read of an angel being used to cause the death of the godless and arrogant King Herod by means of a vile disease (Acts 12:22, 33).

Angels will also be involved in gathering all evil people at the end like weeds, after which they will be delivered to judgement (Matt 13:40–42). In Revelation 14:15–20 John sees in a vision how the ungodly are harvested like vintage and trampled in the wine press of judgement.

7 THE LORD JESUS AND ANGELS

We have already discussed the service which angels perform for and in the name of God. In addition it is essential to pay attention to the particular relationship which exists between the Lord Jesus and the holy angels.

As has already been mentioned, the angels are among others God's messengers to man. These messages also concern, in a special way, Christ's coming to this earth. Luke 1:11–19 describes how the angel Gabriel announced the birth of John, the forerunner of Christ. The same angel also appeared to Mary and informed her that she would be the mother of the Saviour of the world (Luke 1:26–38). Joseph, to whom Mary was betrothed, was reassured by an angel about her pregnancy with the words: " 'for that which is conceived in her is of the Holy Spirit.' " (Matt 1:20).

After Christ's birth in Bethlehem, an angel conveyed the good tidings to the shepherds. On this same occasion we also read: "And suddenly there was with the angel a multitude of the heavenly host praising God and saying: Glory to God in the highest, and on earth peace, goodwill toward men!' " (Luke 2:8–15.)

When Herod planned to murder the Christ-child, Joseph was warned by

26 On the ground that the angels will accompany Christ at his second coming and in view of what we read in Jude, Gordon Lindsay remarks: "Each angel perhaps will be assigned the task of raising the body of one of the saints, as was Michael." *Ministry of Angels*, p 19.

There is, however, nothing in the Bible to warrant the idea that angels will raise people from the dead. With reference to what Jude 9 says about the dispute between Michael and the devil over Moses' body, this is something that belongs to the past and can have no bearing on the day of resurrection.

an angel and commanded to flee to Egypt (Matt 2:13–14). After the death of this king, Joseph was commanded by an angel of the Lord to return to the land of Israel (Matt 2:19–20).

We also read how the angels ministered to the Lord Jesus when He lived on earth in human form. One such occasion was after He had been tempted by the devil. We read in Matthew 4:11: "Then the devil left Him, and behold, angels came and ministered to Him." Another instance was after Christ's struggle in Gethsemane. We are informed in Luke 22:43: "Then an angel appeared to Him from heaven, strengthening Him." Should He have needed any assistance from heaven, He could immediately have claimed the support of twelve legions of angels (Matt 26:53).

We also notice how Christ links his earthly ministry to the revelation of angels (John 1:51). He says to Nathaniel: " 'Most assuredly, I say to you, hereafter you shall see heaven open, and the angels of God ascending and descending upon the Son of Man.' " By this phrase, *opened heavens* we should understand that the invisible things, the heavenly realities or revelations of God, would now become visible in Christ's ministry.[27]

We also see the part played by angels at the resurrection of Christ. An angel of the Lord rolled away the stone in front of the tomb. Hereafter the angel reassured the women who had arrived there and told them that Christ had risen. They were to go and convey this message to the disciples (Matt 28:1–7). Luke 24:3–7 mentions two men in shining clothes who were in the empty tomb and announced the resurrection of Christ.

At the ascension of Christ were two angels – described as men in white clothes – who said to the disciples: "This same Jesus, who was taken up from you into heaven, will so come in like manner as you saw Him go into heaven" (Acts 1:10–11).[28]

While Christ is in heaven, we read that " angels and authorities and powers have been made subject to him" (1 Pet 3:22). This is a reflection of what Christ meant in Matthew 28:18: " 'All authority has been given to Me in heaven and on earth.' " In heaven he is honoured and glorified as the Lamb of God by angels (Rev 5:11–12).

At the return of Christ holy angels will appear with him (Matt 16:27; 25:31; 1 Thess 1:7). Angels will also be present on the day of judgement

27 In this respect L Morris quotes the following statement of Strachan: "The wide open heaven and the ascending and descending angels symbolize the whole power and love of God, now available for men, in the Son of man." *Commentary on the Gospel of John,* p 171.

Matthew Henry interprets it thus: "Through Christ we have communion and benefit by the holy angels, and things in heaven and things on earth are reconciled and gathered together. Christ is to us as Jacob's ladder (Gen 28:12), by whom angels continually ascend and descend for the good of the saints." *Matthew Henry's Commentary on the whole Bible.* Vol V, p 870.

28 About the *cloud* that received Christ and removed Him from the sight of his disciples, Billy Graham writes: "Jesus had been accompanied to earth by an angelic host. I believe that the word "cloud " suggests that angels had come to escort Him back to the right hand of God the Father." Op cit, p 123.

Man may associate the heavenly *cloud* with angels. That implies that the words of Revelation 1:7: "Behold, He is coming with the clouds," and Matthew 25:31: "When the Son of Man comes in His glory, and all the holy angels with Him," must be seen as a consonant description of the second coming.

Cloud in this respect may also refer to God's glory in Nehemiah 16:42; Deuteronomy 5:22. This is apparently also the meaning of cloud used in Revelation 10:1: "And I saw another mighty angel coming down from heaven, clothed with a cloud."

when Christ will judge mankind in the presence of his Father (Mark 8:38).

8 THE HOLY SPIRIT AND ANGELS

It is a fact that angels are spirits who go forth from God and then act on behalf of the chosen. However, the question also arises: What is the relationship between the angels and the Spirit of God (also called the Holy Spirit)?

In this connection we point out certain truths directly stated in the Bible or which can be deduced from it.

As a person within the Trinity, the Holy Spirit is God himself (Matt 28:19). In contrast angels are creatures created by God and fellow-servants of people (Rev 22:9). Therefore angels are not divine and must not be honoured like gods.

Whilst the Holy Spirit is eternal (Heb 9:14), omnipresent (Ps 139:7; Acts 2:17) and so on, angels exist as creatures (Col 1:16) with accompanying limitations. The special powers which angels do reveal are granted them by God.

Angels can in fact communicate something of God to man and create certain conditions which promote the kingdom of God. However, it is the Holy Spirit who can effect God's revelation in a person. Angels can never substitute the Holy Spirit.

Angels can indeed play a part in one's conversion and assist one in living a pious life. However, this help is only of an external nature. Angels cannot convict people of sin, righteousness and judgement. Moreover, they cannot regenerate people nor bring about the fruit of the Spirit in their lives. All this is the work of the Holy Spirit (John 16:6–8; 3:5–6; Gal 5:22).

Angels reveal themselves as beings outside of the human experience. In every angelic appearance recounted in the Bible, angels appear like people who can be seen and spoken to. For this reason it was not always possible to distinguish between people and angels (Heb 13:2). With regard to the Holy Spirit we can only say that he never approached people as an outwardly observable person. The Holy Spirit communicates directly in the mind and the heart of man. Man becomes aware of him, not by sensory observation but by that which he works in man's spirit.

Angels are used as servants who, in an outward way, prepare the way whereby the Holy Spirit can be revealed in people's lives. In Acts 10:30–33 we read of an angel who commanded Cornelius to summon Simon Peter to visit his home. There Peter presented the gospel to the Roman household. They believed and were filled with the Holy Spirit (Acts 10:34–48). This incident as well as the conversion of the Ethiopian court official (Acts 8:26–40) shows us that angels did not proclaim the gospel. Angels can in fact minister in such a way that the preacher and the hearers are brought together. When this happens, the Holy Spirit, by means of the proclaimed word, can work in people's hearts.

One gets the impression that there was a greater need for the part played by angels in the dispensation *before* the outpouring of the Holy Spirit. In earlier times visible phenomena such as the appearance of angels were necessary to bring people into contact with God. However, with the coming of the Holy Spirit, Christ could be directly presented to people by the word which was preached. As people became more and more filled with the Spirit and lived and walked in the Spirit, they increasingly had continual, direct access to God himself. Consequently, the Bible also teaches us that the Holy Spirit came to lead us into the whole truth and to teach us everything about Christ (John 14:26; 16:13–14). Therefore it is understandable that angels are no longer as necessary as before to convey God's will and message.[29]

9 ANGELS AND PEOPLE OF TODAY

In the light of the preceding discussion, the question arises whether we can still expect the appearances and operations of angels today.

At the outset, we should point out that nothing said in the Bible excludes such a possibility. If God has determined in the past that angels should appear to people, then He can still do so today. Consequently there are people who claim that angels appeared to them or that angels did certain things they know of.

9.1 The so-called guardian angels

One reason why people expect angels still to appear to us today, is the view that an angel or angels are assigned to every person. These angels would then be assigned to protect people and help them with that which they cannot do themselves. These angels are known as guardian angels.

This view of guardian angels is based on what the Lord Jesus said in Matthew 18:10 in connection with little children: "Take heed that you do not despise one of these little ones, for I say to you that in heaven their angels always see the face of My Father who is in heaven."

Another Scripture passage which is connected with this view of guardian angels is found in Acts 12:15. After his miraculous rescue from prison, Peter knocked on the door of the house where a number of believers were assembled. When the maid-servant reported that Peter was at the door, they said: "It is his angel." An event like this one, however, has nothing to do with the actions of an angel. It reflects a superstition prevalent among the Jews.[30]

Linked to this, we also have to point out that guardian angels are mentioned in the Talmud and certain extra-canonical writings.[31]

To begin with, we need to obtain clarity as to how we should understand Matthew 18:10. There are different opinions about this passage.

Some, such as Gordon Lindsay, interpret Christ's words as meaning that every child receives a guardian angel at birth. This angel would then have the specific assignment to guard this child for the child's own sake. Further-

29 Concerning this matter, Jessie Penn–Lewis remarks: "A careful study of the epistles of Paul seems to make it clear that God, having 'spoken to us in His Son', no longer speaks by His own direct voice to His people. Nor does it appear that since the coming of the Holy Spirit to guide the Church of Christ in all the truth, does He frequently employ angels to speak or to guide His Children." *War on the Saints*, 1983, p 78.

30 In this connection Gerhad Kittel remarks: "The idea of a guardian, or better the directing and ministering angel, is taken over from Judaism, which had long since forgotten the animistic roots of the notion. Acts 12:15 assumes a likeness in appearance and voice between the 'angelos' and the man concerned." *Theol Dict NT*, 1, p 86.

31 According to the Talmud 11 000 guardian angels are assigned for each Jew as long as he lives. Mention of faith in guardian angels is found in The Book of Jubilees, 35:17, and also in Philo's The *Biblical Antiquities*. Cf Gustav Davidson, op cit, p 128.

more, the angel's task would include taking the spirits of deceased children to the paradise for babies. There the angel would instruct the little ones and develop their natural talents. (This is based on the vision of a certain Marriëtta Davis who was in heaven for nine days.)[32] Thus the *their angels* in Matthew 18:10 would refer to the spirits of deceased infants. The fact that these spirits or angels gaze upon the face of God implies that they are saved. Others interpret Matthew 18:10 in conjunction with Acts 12:15. The notion is that Peter was put to death and that his spirit (angel) knocked at the door.[33]

These statements necessitate some comments. The Lord Jesus is not in this instance speaking about the condition of little children after death. He is reprimanding his disciples to ensure that these little ones are not despised. He does not give this reprimand on the grounds of what these little ones would be if they were to die, but on account of the fact that their angels now see the face of the heavenly Father. Besides, we have pointed out that what we read in Acts 12:15 refers to a superstition prevalent in Judaism at the time.

Others maintain that the *little ones* mentioned here refer to the disciples of Jesus.[34] However, considering the context in which *little ones* is mentioned here, it is impossible to interpret it as a reference to the disciples. After all, it is unthinkable that the Lord Jesus, speaking to his disciples (cf Matt 18:1), would refer to them in the third person, namely as *such little children* and *these little children.* Referring to the disciples' concern about who is the greatest in the kingdom of heaven, the Lord brought a small child into their midst. He then began to speak about small children and cautioned the disciples not to despise them or allow them to stumble. So how can the disciples themselves be these little children?

Matthew 18:10 confirms the special significance which children have for Jesus by stating that their angels appear before the heavenly Father on their behalf. What this means is not explained. All we can say is that the angels are not interceding for the little children. Nowhere do we read that such a ministry has been assigned to angels.[35] These angels also do not act as Mediator and Advocate on behalf of people before God. Christ alone is our Mediator and Advocate. To see the face of the heavenly Father simply means that the angels are in a direct relationship with God.

While this passage is concerned with the status of little children in the eyes of the Lord, we also have to pay attention to what He says concerning them in Mark 10:14, namely: " 'Let the little children come to me, and do not forbid them; for of such is the kingdom of God'." The fact that the Lord makes this gracious pronouncement concerning little children presupposes that He will also take special care of them. This special care apparently also includes the assignments which angels are given on behalf of the

32 Gordon Lindsay, op cit, pp 6, 7.

33 Guy P Duffield and Nathaniel M van Cleave, op cit, pp 476, 477.

34 Cf F W Grosheide, *Het heilige evangelie volgens Matthews*, 1954, p 278, as well as J Rodman Wiliams, *Renewal Theology*, 1, p 191.

35 About this matter K Dijk writes: "About separate guardian angels, one for every believer, holy Scripture knows nothing, even though God might assign a specific task to a specific group (their angels, Matthew 18:10), and at the same time there is nothing in God's Word about their intercession on our behalf in heaven or our worshipping them on earth." *Korte Dogmatiek*, p 58.

little children. Notice that the reference here is not to a single so-called guardian angel. (Cf also Ps 91:11.) Moreover, if we keep in mind that it is the function of the angels to combat evil spirits, we can assume that they are assigned to protect little children from seduction by the Evil One. This idea links up with what we read in verses 12 and 13 of the same passage (Matt 18). If a shepherd is concerned about a single sheep which has gone astray, how much more will the Lord concern himself with little ones. Therefore this section is also concluded with the words: "Thus it is not the will of your Father who is in heaven, that any one of these little ones should perish." (Matt 18:11 does not appear in certain important manuscripts. Thus it is uncertain if this passage was originally included in the Gospel of Matthew).

Whatever interpretation we may append to Matthew 18:10, one thing is, however, certain: we have no reason to develop a theology of a so-called guardian angel for each person.

9.2 Angels in the service of man

We have already pointed out that Hebrews 1:13–14 speaks of angels sent as ministering spirits on behalf of those who will receive salvation. In connection with this the following question arises: Can a person do something in order to claim this ministry by angels?

Some answer this question in the affirmative. They maintain that the way to get the angels to work on their behalf, is to apply a promise from the Bible, according to the natural law of cause and effect.

In this connection we refer to the statements by Charles and Annette Capps. They maintain that there are angels with each one of us, similar to the so-called guardian angels. These angels are merely waiting for people to give them the correct instructions. If you eagerly want to see a biblical promise fulfilled in connection with your problem, then you should merely follow the correct procedure. You should then say the appropriate verse aloud so that the angels can hear it. At the same time you should believe that the Lord will cause these heavenly beings to work for you now. If you follow this procedure, that is, if you put this mechanism into operation, you free the angels to carry out your desire. Should you, however, say something negative, such as for example that a certain matter is hopeless, then you bind the angels and they can do nothing for you.[36]

The Capps couple also maintain that angels do not work for a special purpose, such as the salvation of mankind. Instead they are like spiritual machines which merely do what they have been commanded. When they receive a positive message, they immediately jump to action. Assignments to angels are likened to the kind of instructions given in an emergency. It also includes assistance to find things which are lost. They can even lock your car door should you experience difficulty when doing this. In emergencies angels can also repair your car. Such repairs would, however, only last until one can fix it in a normal way.[37] The words of Paul in 1 Corinthians 6:3, namely "Do you not know that we shall judge angels?" are also quoted

36 Cf Charles Capps and Annette Capps, *Angels*, Harrison House, Inc. Tulsa. pp 75–90, 209.
37 Charles Capps and Annette Capps, op cit, pp 132, 177–182.

in this connection. In this way we can hold the angels to account, by for instance asking them: "Why did you allow me to have an accident?" [38]

This view represented by the Capps supposes that if you confess Bible verses with faith, it becomes the angels' responsibility to ensure that things work out for you. Thus all problems can be solved provided you make use of the correct spiritual mechanisms! The Bible (as well as practical life) teaches, however, that this is not the way these things work. For example consider Job, Psalm 73, Paul's experiences (2 Cor 11:24–28) and numerous other Scripture references where true believers had to wrestle with problems without angels or any other party protecting them from these trials.

Indeed, God does want us to believe in his promises, but by believing we do not take the initiative away from God and his angels. Faith is not a procedure whereby man can compel God and his angels to solve his problems and protect him from all kinds of dangers in a simple scheme of cause-and-effect. Faith is not a kind of switch to activate a certain mechanism whereby one obtains the desired results. Faith implies standing in a relationship of complete trust in God. In that relationship the believer can trust in God for everything according to his promises. However, God fulfills these promises in a way and at a time decided by Him. God's care for believers does indeed include instances where things turn out exactly the way people entreated Him. It is also possible that angels may be operative in such answers to prayer, even though man himself may not be aware of this.

In all the instances in the Bible where angels were used to rescue people from grave dangers or where insurmountable problems were solved, it happened in a way that man neither expected nor prescribed. The person involved was merely granted the opportunity to testify of angelic help.

It must be a source of comfort to know that angels are used by God for our benefit. However, we cannot relegate all answers to prayer to angelic agency such as is taught by the Capps. One can only say that angels were involved when this had been revealed to us.

9.3 Contemporary appearances by angels

From time to time we hear of people who claim that an angel appeared to them. There are numerous instances where a motorist on a journey suddenly became aware of someone in the passenger seat next to him. The latter would then express some thoughts about spiritual matters, mainly about the imminent return of Christ. Thereafter the person would disappear into thin air in the same way in which he came.

A woman once recounted that while her pious old mother was dying, an angel appeared in the corner of the room. After the mother's death, the angel disappeared. The idea behind this is that the angel came to fetch her.

A girl once went swimming on her own and was carried into the sea by the current. After she had given up struggling and began to lose all hope of being saved, she suddenly called out: "Lord help me!" The following moment someone on a small island stretched out his hand to her and

38 Charles Capps and Annette Capps, op cit, pp 129–132.

brought her safely onto a rock. After she had rested for a while, the person pointed out the direction in which she should swim. While swimming to the land, she looked back and waved to the man on the small island. He waved back at her. After arriving on the beach she contacted her friends and told them of the man on the small island who had rescued her. "What island?" they asked in surprise. When she looked back, there was no island in sight. Her rescuer had also disappeared. In all probability it was an angel who had appeared and rescued her.

The question now arises whether these stories indeed involve angelic appearances. In response we have to state at the outset that no-one has the right to dismiss summarily all accounts of this nature as mere flights of imagination or hallucinations. However, we should test appearances ascribed to angels in the light of Holy Scripture. If that which a person claims were revealed to him by the Spirit of God has to be tested (1 Cor 14:29; 1 John 4:1–3), how much more should claims of angelic appearances be subjected to testing! If this is done, we find that some reports about the appearance and actions of angels are seriously suspect.[39] Besides, we should also not dismiss the possibility of satanic deception. Paul also warns us in 2 Corinthians 11:4 that Satan can appear in the form of an angel of light.

9.4 Are angels aware of what is happening to people on earth?

The above discussion now poses the question whether angels are in any way aware of what is happening to us here on earth.

Certain statements in Scripture indeed show that angels see and hear what people are doing and know what people experience on earth.

For angels to be able to minister to those who will inherit salvation (Heb 1:14) implies that they must have knowledge about such people. This knowledge can be obtained by means of information which they get from God, or by means of their own observation or both.

In the Bible we read about an angel who accompanied Abraham's servant on a journey and who knew everything that this man was about to do (Gen 24:7. Cf also Exod 23:23).

Angels knew what took place in the land of Israel. For this reason an

39 As an example in this regard, we refer to Roland Buck who avowed that over a period of two years numerous angels appeared to him and they conversed altogether about fifty hours.

Some of the things the angels communicated to him can already be read in the Bible, or are general knowledge scientifically speaking. These angels supplemented events of the Old Testament in ways which are strange and unacceptable. Examples are the statement that the walls of Jericho did not fall down flat (Josh 6:20), but angels pushed them into the ground, and that angels formed balls of ice with which they routed the Amorites (Josh 10:11).

He also told how the angels helped him to reduce weight in order to jog more easily. Angels even played with his dog which enjoyed it immensely. On a certain occasion they even used the ground floor of his house as headquarters in their battle against evil angels. Their consultations were so rowdy, that Buck had to go downstairs to investigate.

There were conflicting communications from the angels regarding who the Commander of the army of the Lord was that stood opposite Joshua. He also reports how angels prevented a typhoon from destroying a Philippine town, yet allowed it to devastate the rice fields!

Likewise there are other reports which are not only platitudinous, but which are not reconcilable with the loftiness of the biblical angel appearances. All these are written in the book: *Angels on Assignment* by Charles and Frances Hunter, as told by Ronald Buck, 1979.

angel could tell Joseph that King Herod had died, and that Joseph, together with Mary and the Child should return from Egypt to their native land (Matt 2:19–20).

We read that angels know and rejoice when a sinner on earth repents (Luke 15:10–15).

In 1 Corinthians 4:9 Paul mentions that believers have been made a spectacle to both men and angels.

According to 1 Corinthians 11:10 angels know whether women have an indication of authority on their heads.

In 1 Timothy 5:21 Paul says that his commands to Timothy were given in the presence of God, the Lord Jesus Christ and the chosen angels. It is before these witnesses that he then calls Timothy to account.

Romans 8:38 implies that evil angels are so involved with people that they can be described as a potential danger which may separate us from God. How much more can we expect the holy angels to involve themselves in protecting us in our relationship with God!

The congregation, namely the assembly of believers, is able to make angels aware of matters regarding God (Eph 3:10). This is only possible if angels are able to observe people's deeds on earth.

10 CONCLUDING REMARKS

Since angels are spirits directly under God's control, we cannot say anything more about this topic than that which has been revealed to us in Scripture. We must not construct an angelology from practice and human experience.

When we talk of the possibility of contemporary angelic appearances, we have to consider the period in which we live. In the past God did in fact speak in many ways to people, yes, even by means of angelic appearances. In these last days, however, He speaks to us through His son (Heb 1:1). We are now dealing with the gospel of Christ. Keeping this in mind, it is to be expected that angelic ministries are no longer intended to convey God's message to people. Instead we can expect angels to serve the Kingdom of God in other ways. For instance, we think of the protection of believers and the creation of circumstances conducive to the proclamation of the gospel. We may also expect angels to act less conspicuously in human circumstances and history precisely in order to promote Christ and his gospel.

In the hereafter, when we shall be fully informed, we shall also learn more about the role which angels played in our lives. (1 Cor 13:12)

Man is not in a direct or personal relationship with angels. His relationship is with God and only through Him does man deal with angels.

We cannot, by ourselves, determine whether we are dealing with angels or not. We can only know this if it is specifically revealed to us or if the nature of a miracle is such that the operation of angels can be deduced from it. This deduction will, however, remain a matter of faith.

I believe that when we ask God to change our outward circumstances or rescue us from a situation of danger, He may use angels for this purpose. Such has also been the testimony of various people whose word we have no reason to doubt.[40]

If God were to give a particular instruction or message by means of an

40 We refer to the incidents recorded by Billy Graham. Op cit, pp 13–15, 154, 155.

angel to someone regarding the church, we believe that He would do it in a way which believers would find acceptable. It would, for example, be more acceptable if this revelation came to a dedicated leader instead of anyone else. That was the manner in which the Lord made his mind known to the seven churches in Asia Minor (Rev 2–3).

The existence of angels reminds us to focus our attention also on the actual reality, that is to know God who is Spirit and the things of the spiritual realm. Allusions to the superiority of spiritual things over material things, is expressed throughout Scripture. In Isaiah 31:3 Israel is admonished not to accept the futile help from Egypt: "Now the Egyptians are men, and not God; and their horses are flesh, and not spirit, when the Lord stretches out his hand, both he who helps will fall, and he who is helped will fall down; they all perish together."

As has already been said, it is a source of comfort and strength to know that God sends angels as ministering spirits for the sake of those who will inherit salvation. As believers, we are never alone. The Lord has promised to be with us for ever, through his spirit (Matt 28:20; Rom 8:38–39). Therefore, we can accept that the angels who stand in God's presence and carry out his instructions may also be with us, yes, even if we cannot see them.[41]

The knowledge that angels see and hear what we do and say here on earth, should be an incentive to sanctification and service.[42]

41 With regard to this, J Rodman Williams remarks: "The emphasis is wrongly placed when the focus is on angelic visitation; indeed, expecting for, looking for, or hoping for such visitors is nowhere encouraged in God's World. We are rather to pray and expect, especially in our day, a greater visitation of the Holy Spirit (that's where the action is!). And as far as angels are concerned, we may rejoice in their invisible but continuing providential presence." Op cit, p 195.
42 Cf Billy Graham, op cit, pp 145, 146.

CHAPTER 6 Demonology: Doctrine about Satan and his angels

In the previous chapter the fall of a great number of angels was mentioned. Thus, in addition to the chosen and holy angels, we also deal with rejected and evil angels. This brings us to the doctrine of demons or the teaching about Satan and his angels.

The belief in the existence of the devil and evil spirits more or less followed the same course as the faith in angels in general. Man's struggle with invisible evil forces forms part of the cultural history of every nation on earth. This is especially true of very ancient civilisations, but also occurs in contemporary primitive civilisations where much attention is given to evil spirits and man's struggle against them.

In biblical times the existence of the devil and evil spirits was acknowledged as a reality which the believer had to battle against. There were even times when the church showed an excessive interest in and even a dangerous preoccupation with demonic activities. Here we think in particular of the Middle Ages when all kinds of superstition and witchcraft occurred. Thus the 13 th century is typified as the *"devil period"*. For this reason Pope Innocentius VIII was forced to promulgate the infamous bill *Summis Desiderantes* in 1484. According to this, procedures against witches and their burning were ecclesiologically sanctioned.

With the dawn of the so-called Enlightenment in the 17 th century, reason triumphed over every form of superstition. Together with the exaltation of reason, belief in evil spirits (yes, even in holy angels) was rejected as belonging to the world of make-believe.

Different theologians included this thinking in their interpretations of the New Testament. According to Bultmann, the Bible should be demythologised of notions of demons and Satan. Accounts where Christ (and his apostles) cast out demons would show that he had adapted himself to contemporary popular belief or even endorsed it. Therefore, he would also have considered psychiatric patients and certain physical illnesses as cases of demon possession. The deliverance experienced when demons were cast out of people should instead be ascribed to the power of suggestion whereby psychological bondages were dissolved. What has been described as demons was in reality merely sinful tendencies. Exorcism would then only comprise the exposure of these things by preachers who would then convince people to abandon them. In particular, we find this view among adherents of liberal theology from the 18 th and 19 th century.

In response to the above statements, we make certain comments. The truth is that there is no proof of a general existence of such popular belief in Israel during New Testament times. The Sadducees did not even believe in demons at all, while certain rabbinical writings portrayed them in a completely different light to that which we read in the Gospels. Accordingly we can ask: if the aspect of Christ's ministry where He exorcised demons is false, how do we know if anything He did and taught is true?[1]

Influenced by the positivistic scientific ideal of the 19 th century, psychology interpreted all faith in God and belief in devils as human projections. This was seen as the actualisation of man's need for a father figure (in the case of God) and his fear of threatening evil forces (in the case of the devil and evil spirits). This view continued right into the 20 th century.

However, events in the 20 th century have once again led people to start talking more about the devil and evil spirits. The gruesome events which took place during the First, but especially the Second, World War, were described as demonic. Subsequently the church had more to do with the occult against which it had to assume a position. All kinds of dangerous practices were encountered on the mission field, demonisation and exorcism occurred and there was a revival of satanism and related matters. Therefore, it is not surprising that many books (and even films) about the devil and demonic operations appeared.

The so-called scientific and rational viewpoint which dismisses all supernatural phenomena as *medieval* and *anachronistic* is in reality not a sign of expanded and more sober thinking. Instead it is a materialistic and mechanistic narrowing of reality.[2] We dare not reject the biblical teaching about Satan and demons just because erroneous ideas and practices surrounded this topic in the past.

There are theologians, such as G C van Niftrik, who maintain that demonology does not belong in dogmatics. He asserts that discussing demons, while studying matters related to God and the holy angels, means that too much attention is given to the former. In addition it is alleged that the devil has no place in that which God has willed and created. For this reason evil should be given no place in dogmatics, since the latter should concern itself only with God's acts of salvation. One should exercise theological exorcism. This implies a radical unbelief in the devil and his cronies.[3]

However, such a view cannot be maintained. One cannot speak of Christ and his work of salvation or of man and his struggle against evil and sin, without paying attention to the devil and his angels. In any case if the Bible

1 Cf W C van Dam, *Demonen eruit, in Jezus' Naam*, 1974, pp 45–53.

2 Van Dam says the following: "In larger circles, especially amongst the youth, the elevation of reason had to make room for the search of broadening the mind. Many are groping for spiritual realities outside the scope of cognitive consciousness. Mysticism, meditation, drugs and occultism are employed to open up the way to this. With this, one is unavoidably confronted with the reality of evil spirits. Satan worship is springing up everywhere. That which many considered to be medieval and anachronistic, has become very actual anew. The New Testament's vision of the struggle between the Kingdom of God and the kingdom of darkness, has become again very close to us ... Critics who call demon exorcism anachronistic, are themselves rapidly becoming anachronistic." Op cit, pp 51, 52, 54.

3 G C van Niftrik, *Kleine Dogmatiek*, pp 104, 105.

says a lot about evil angels, these pronouncements must also be dealt with in dogmatics.

1 THE EXISTENCE OF EVIL ANGELS

The Bible does not endeavour to prove the existence of Satan and evil spirits. The existence of these beings as a sinful and hostile reality is accepted without any ado.

1.1 The rejection of the existence of evil angels

As has already been mentioned in the discussion about angels, certain theologians relegate the existence of the devil and evil spirits to mere myths. At best, some regard this as the personification of superhuman destructive systems.[4]

One reason why critically inclined people do not believe in the devil and evil spirits is because of the warped and unbiblical perceptions about evil forces currently in vogue. The devil, for example, is portrayed as a hairy human-like creature with the split hooves and horns of a goat and a pointed tail. To many others evil spirits imply things like ghosts with which you scare children and credulous people.

Something else which indeed contributed to the rejection of the belief in the existence of evil angels, was the strange ideas and practices regarding demons which occurred in the medieval church. As we have already said at the beginning of the chapter, this was the time when the church was obsessed with hunting down and burning witches. If someone did something unusual or spoke out against existing ecclesiological ideas and customs, it was forthwith considered demonic. Consequently it seems that no questions of conscience could stop the very cruel persecution of these people.

As has also been mentioned in a previous chapter, evolutionism obviously also rejects the existence of evil angels. The origin of social evils, struggle, disorder, suffering and so on is not attributed to a superhuman power; it is merely a stage in the process of development to a higher and better form of existence. With the necessary further evolutionary development all such so-called sinful and evil forces would in fact be eliminated!

Apart from the fact that such a viewpoint is completely unbiblical, it is also self-condemnatory. Evil, as experienced on earth, is not the preamble to better things or the generator thereof. Nothing good emerges when hatred, cruelty, deceit, anarchy, meanness and reckless destruction of all that exists is the order of the day. Such things should be combated forthwith and eliminated if a better dispensation for people is expected.

We have to accept that, as far as man is concerned, it fits the devil's strategy nicely to remain unknown and anonymous. As long as he is not identified as an evil force, it is easier for him to continue with his work of destruction. Therefore the devil does not want himself to be known as an agent outside of or apart from man. He would much rather prefer things

4 Cf R Bultmann, *Neues Testament und Mythologie*, 1941. P Tillich, *Systematic Theology II*, 1968, pp 41–50.

directly connected to him to be viewed as the product of human thinking, state of mind, maladaption or something similar.[5]

1.2 The existence of evil spirits cannot be separated from the biblical message

The fact is that one cannot believe in the truth of the Bible without acknowledging the existence of the devil and his angels. The existence and deeds of evil forces are mentioned everywhere in the Old and New Testaments. The coming of Christ to this earth, his work of reconciliation and the Gospel would be senseless if there were no Evil One or evil forces from which man has to be delivered. Stories where the Lord Jesus and the devil converse, or where the Lord drove out demons, then would not have occurred in the Bible. The Bible expressly teaches that the Lord came to break the power of the devil (1 John 3:8).[6] Thus a common theme in the Holy Scripture is that of admonishing believers to take a stand against the devil and overcome him through Christ. A book such as Revelation can only have significance and be understood if we believe in supernatural forces which exert an influence here on earth. This force will ultimately be destroyed by Christ at his second coming.

We should pay attention to the fact that this evil force is not something impersonal as maintained by certain theologians.[7] The Bible refers to Satan and the evil spirits in a personalised way. The devil is described as the chief or prince of this world, the father of unbelievers, someone who can will, speak, plan, seduce, deceive, has knowledge, etc. Similarly, evil spirits are described as beings who can speak, do supernatural things through people and who are destined for judgement.[8]

However, we have to point out that the devil and his angels do not comprise a separate species which exists beside other angelic beings. As we have already mentioned in the chapter dealing with sin, evil is not a being in

5 Regarding this, we refer to some observations in JA Heyns' summary of *La Part du Diable* by Denis de Rougemont. "The existence of the devil is, however, not implied on theological philosophical grounds. It can be deduced in an illustrative manner from the conduct and thought of people. The devil operates behind masks; he performs in disguise. It is therefore the task of the believer to unmask him – in the thoughts and life of other people, structures and institutions, but also in his own life and thought. He goes about and works incognito, and that is what must be exposed. He who does not believe it, and he who does not want to do it, finds himself in the clutches of the devil." *Ned Geref Teologiese Tydskrif*, September 1975, p 352.

6 In this respect K S Latourette states: "He (Jesus) saw his healings, especially the cure of those troubled by demon possession, as evidence that it (kingdom) had come. To him the forces of evil were intensely real and personal. He accepted Satan as an existent being, an enemy of God, and he addressed the demons whom he cast out, not as delusions of sick minds, but as actual." *A History of Christianity*, 1955, p 41.

7 H Berkhof refers to the Jewish apocalyptic which considers these powers as personal, spiritual beings which have an influence on earthly events, especially in the field of nature. He asserts that Paul demythologises the notion of heavenly beings to unpersonal, earthly relationships of life or ordinances which he just personalises. Thus, as an example, the expression 'rulers of this age' (1 Cor 2:6) must be associated with evil institutions, like that of Nazi-Germany. *Christus en de Machten*, 1952, pp 7–65.

8 In this connection E Sauer writes: "The existence of sorrow and evil through the world proves the existence of a transcendental, real, dynamic, hostile power, not willed by God. This must be thought of as a personality, since the ethical and unethical can only be predicated of a conscious, free being." Op cit, p 62.

itself; it should rather be regarded as the sinful deeds manifested by creatures. This sinful nature of their life and behaviour points to that which is anti-God, anti-man, anti-creation, anti-life, anti-good, anti-truth, anti-order, anti-beauty, etc. Without a prior creation which was good and perfect in all respects, there can be no question of an anti-force. Consequently we can regard evil as a foreign and parasitic force revealed by creatures who were formerly good.

1.3 The origin of evil spirits

At this juncture we can pose the following question: If God created everything good, where then do evil beings come from?

As has already been mentioned, not only did God create everything good, He also gave angels and people a free will. With this the possibility was created that creatures such as angels could act against God and his will, degenerate and fall under judgement. The tragic fact is that this actually happened. The angels (and later also man) rebelled against God and thus acquired an evil nature. Although they consequently became evil angels, God, nevertheless, did not destroy them immediately; they are still sustained as his creatures.

However, if this was all we could deduce from the Bible, we would be left in total confusion. Then we could indeed ask: How can God, who is good, allow the perpetuation of evil?

We can only begin to understand the answer to this question when we pay attention to the end of all things. We see that everything which is and which has taken place, leads to one climax, namely the triumph of Christ. This triumph implies that not only the believers, but also the evil forces will eventually be subjected to Him (Phil 2:9–11; 1 Cor 15:25–28). As has been mentioned already, Christ triumphs over and ultimately destroys the Evil One, together with the consequences of all evil. At the same time demonic powers cause the battle in which the believers are involved (1 Tim 6:12), and by which they, together with Christ, and the grace of God, emerge as victors. Together with this the opportunity is created in whom the love, grace and mercy of God can be revealed toward man in a special way. In this process of battle and victory, man develops into the being whom God intended. Above all, and the ultimate end to which everything leads, is that God is glorified through Jesus Christ and the believers.[9]

2 THE CLASSIFICATION: SATAN AND DEMONS

Evil angels are usually categorised in two groups: Satan or the devil on the one hand and demons or evil spirits on the other. Satan is the head of the demons who are known as his angels (Matt 25:41; 2 Cor 12:7; Rev 12:7, 9).

2.1 The different references to Satan

We see something of who Satan is and what he does in the different ways in which reference is made to this chief of demons.

9 Martin Luther declares that God employs Satan (and the evil angels) for his alien work (opus alienum) which is subservient to his actual work *(opus proprium). Werke* 56, 331, 27.

2.1.1 Satan

Satan is the fallen Lucifer (Morning Star). *Lucifer* is a composite of two Latin words which can be translated as *light bearer* or *light bringer*. Lucifer's name therefore points to the glorious state in which this angel existed before he fell.

The word *Satan* in Hebrew, Aramaic and Greek literally means *adversary*. The word *Satan* is even used separately from the concept of the devil. It can mean adversary *per se*. Thus the Angel of the Lord explains (Neh 22:32) that he acted as Balaam's adversary (satan). In 1 Samuel 29:4 the Philistines regard David as a possible adversary (satan).

Satan, as the devil, comes before God as Job's adversary (Job 1:6–12; 2:1–5) whilst in the case of the high priest, Joshua, he is again the accuser (Zech 3:1–2). Satan is also known as the tempter, deceiver and seducer (Matt 4:10; Acts 5:3; 2 Cor 11:14; Rev 12:9).

The place where idolaters or those who blaspheme Christ are gathered is described as the synagogue or throne of Satan (Rev 2:9; 3:13).

2.1.2 The devil

The Greek word *diabolos* (from which the word *devil* is derived) literally means: accuser, speaker of evil (1 Tim 3:11; 2 Tim 3:3). The verb form of *diabolos* means to *accuse* as used in Luke 16:1.

The Hebrew equivalent for *devil* is *sair* which literally means *the hairy one* or *goat* (Lev 17:7) as well as *shed* which means *destroyer* (Deut 32:17). The word devil is a synonym for *Satan* (or of *the adversary*) (Matt 4:1; Acts 10:38; Eph 4:27; 1 Pet 5:18).

When the plural form *devil* occurs in different Bible translations, it does not refer to the devil or Satan but to demons.

2.1.3 Beelzebub

There is uncertainty about the orthography (way of writing) and the meaning of the Greek word *Beelsebub* as it appears in the New Testament.[10] Webster understands the English word *Beelzebub* as a composition of the Hebrew words *baal* and *zebub* which respectively mean *lord* (god) and *fly*.[11] We can therefore speak of the god of the flies.

On account of what the Jews say, the Lord Jesus describes Beelzebub as Satan and the chief of the devils (Mark 3:22; Matt 10:25).

2.1.4 Belial

Satan or the devil is also called *belial* (translation of the Greek *beliar*). This term is only found in one place in the New Testament. In 2 Corinthians 6:15 we read: "What accord has Christ with Belial? Or what part has a believer with an unbeliever?"

In order to determine the possible meaning of this word we look at a homophonic word in the Hebrew of the Old Testament as it occurs in 2 Samuel 22:5 and Psalm 18:5. In the Afrikaans version of the Bible, it is

10 Cf Werner Foerster, *Theol Dict NT 1*, p 605.
11 The Consolidated Webster Encyclopedic Dictionary, 1947, p 63.

translated with *streams of evil*, whilst the authorised version speaks of *floods of ungodly men*. The essence of the word *belial* as used in connection with the devil is apparently *evildoer*.[12]

2.1.5 Descriptive names for the devil or Satan

Besides the above-mentioned names reference is also made to Satan as: enemy (Matt 13:39); murderer of people (John 8:44); liar (John 8:44); the Evil One (Matt 6:13); Prince of the world (John 12:31); god of this world (2 Cor 4:4); the old snake (Rev 12:9); the great dragon (Rev 12:9); the accuser of the brethren (Rev 12:10); one who poses as an angel of the light (2 Cor 11:14); the father of liars (John 8:44); one who appears as a roaring lion (1 Pet 5:8), etc.

2.2 Demons

According to Holy Scripture, the angels who together with Satan rebelled against God are called demons. It is derived from the Greek word *damonia*, plural of *daimôn*.[13]

Reference is also made to these fallen angels as evil spirits (Luke 7:21), unclean spirits (Matt 10:1) and the messengers of Satan (2 Cor 12:7). When spirits are mentioned here, it should not be confused with the spirits of the dead. The latter are either in paradise (Luke 23:43) or in the place of the dead (Hades) where the ungodly are gathered (Rev 20:13).

Although there is only one Satan or devil, there must be a very large number of demons. For instance in Mark 5:9 a demon-possessed man is called *Legion* because there were so many devils in him. (A legion was a Roman cohort which consisted of between five and six thousand soldiers. However here the name *Legion* does not refer to a specific number but merely indicates a multitude.) After the Lord Jesus had cast out the unclean spirits or demons from this man, they took possession of approximately two thousand pigs which stormed into the sea. Another indication of the great multitude of demons can be deduced from John's vision of the locusts which rose from the abyss under the command of the angel of the abyss, Abaddon or Apollyon. Everything indicates that the beings which look like locusts are in reality evil spirits (Rev 9:1–11).

Since the demons are so numerous, the devil can attack every person on earth with them.

Regarding the demons we read that they can fear (Luke 8:31), speak (Matt 8:28–29), have desires (Matt 8:31), show exceptional powers (Mark 5:4), are aggressive (Matt 8:28), indwell the bodies of people and animals (Matt 12:43–45; Mark 5:12–13), attack believers (2 Cor 12:7) etc.

The demons are divided into a kingdom or an army as is indicated by the mention of authorities, forces, world rulers (Eph 6:12), princes (Dan 10:13, 20) and so on.

12 Cf Werner Foester, op cit, p 607.

13 Regarding this word, Werner Foerster writes: "The etymology of *daimôn* is uncertain. The root *dai* is basic, and cf *daiomai*, thought the sense is doubtful. W Porzig is perhaps right in suggesting disruption or rending apart, and therefore in his conception of the daimôn as that which consumes the body." *Theol Dict NT 2*, p 2.

The demons are in the service of Satan and are responsible for many onslaughts on man. Thus we read that demons can cause physical as well as spiritual illness among people (Luke 13:16; Acts 16;18) and that heresies and doctrines of devils are propagated amongst people (1 Tim 4:1).

3 THE AIM AND POWER OF EVIL

In order to understand evil correctly and combat it, we must know its objectives.

3.1 What the devil is trying to attain

We discern the devil's attitude and aim in what is said about his fall. The devil's attitude is expressed in terms of the king of Babylon. Isaiah 14:13 and 14 put it in this way: "For you have said in your heart: I will ascend into heaven, I will exalt my throne above the stars of God; I will also sit on the mount of the congregation on the farthest sides of the north; I will ascend above the heights of the clouds, I will be like the Most High.' "[14]

The devil's inflated arrogance is also described in terms of the king of Tyre. At first he was an exalted and holy cherub who was later cast down on account of his unrighteousness (Ezek 28:12–17).

Satan's aim and policy is to prevent people from believing in and serving God. He also tries to undermine the faith of those who already serve the Lord and to lead them into apostasy. Satan wants his own counter-kingdom. He attains this as far as possible by getting all humanity into the grip of sin. Those who sin and turn their back on God reveal the devil's spirit and attitude. Paul declares that those who have not yet abandoned sin and turned to God, allow themselves to be controlled by the prince of the invisible forces. In such a one the spirit of disobedience is at work against God (Eph 2:1–3). As the children of God are filled with his Spirit, in the same way the ungodly children are of the devil because his spirit is in them. He is the father of the ungodly (Matt 13:38) and all those who reject Christ (John 8:44).

While the believer is in a personal, knowledgeable and loving relationship with God, the situation of the unbeliever and the devil is totally different. As a rule the unbeliever does not even believe in the devil and is also unaware that his spirit is in him. (The different forms of satanism where people purposefully seek contact with the Evil One is the exception here.) This ignorance on behalf of unbelievers does not, however, alter the fact that the devil exercises his power over them (1 John 5:19) and that he is the prince of the entire sinful world (John 12:31). However, this ignorance suits Satan very well. Due to the darkness of ignorance and the wrong attitude toward God, the Evil One can most easily realise his crafty plan with mankind. Whether one believes this or not, if one does not live to the glory of God, one promotes the cause of the Evil One. As long as people reject God and

14 C F Keil and F Delitzsch maintain that the church fathers unjustifiably associated the term Lucifer with the devil as a fallen angel. The term indeed refers to the king of Babylon, about whom we read in Isaiah 14:12–15. They also say: "A retrospective glance is now cast on the self deification of the king of Babylon, in which he was the antitype of the devil and the type of antichrist (Dan 11:36; 2 Thess 2:4), and which had met with its reward." *Commentary on the Old Testament*, Isaiah. p 312.

do not serve him, the devil achieves his purpose. There is no neutral ground with regard to these matters. If one does not enjoy God's loving care, one is under the power of the devil. This does not imply that one is total demonic; however, it does imply that the Evil One has some power over one.

3.2 The power of Evil

The Bible describes Satan as an evil being with great power. In a vision John saw how the great dragon (Satan), equipped the beast who comes up from the sea with such power, authority and special abilities that people cried out: "Who is like the beast? Who is able to make war with him?" (Rev 13:1–4) (Compare similar visions seen by Daniel of the authorities of this world who assumed the form of terrifying animals (Dan 7:1–8.))

In connection with the Anti-Christ we read: "... the coming of the lawless one is according to the working of Satan, with all power, signs, and lying wonders." (2 Thess 2:9). The devil and his angels are so mighty that they will challenge the Archangel and the heavenly hosts (Rev 12:7). According to Jude 9 we see how Michael, the Archangel, regards the devil with a measure of respect.

3.3 The expulsion of Satan from heaven

In Luke 10:17 we read how the 70 sent out by Christ, reported as follows: "Lord even the demons are subject to us in Your name." In the following verse we find these words: "And He said to them, 'I saw Satan fall like lightning from heaven.' "

Everything indicates that this expulsion of Satan from heaven is directly connected with Christ's ministry and work of reconciliation here on earth. What he did here on earth, enables believers to overcome the devil.

After the theophany (the revelation of God) which we read about in John 12:27–31, the Lord Jesus made the following statement: "Now is the judgment of this world; now the ruler of this world will be cast out." We have every reason to believe that this casting out of the ruler of this world is the same as Satan's expulsion from heaven.

In John 16:11 the Lord Jesus declares that the prince of this world has already been judged. He says this in connection with his own work of reconciliation, ascension and the outpouring of the Holy Spirit.

John also saw these events in a vision. There was war between Michael and his angels on the one hand and the dragon (devil) and his angels on the other. The evil forces were overcome and " they did not prevail, nor was a place found for them in heaven any longer. So the great dragon was cast out, that serpent of old, called the Devil and Satan, who deceives the whole world; he was cast to the earth, and his angels were cast out with him." (Rev 12:7–9.) What was thus conveyed to John is that the power and kingship of God through Christ was confirmed. Together with this it also implies that the devil's position as accuser of the believers before God (Zech 3:1) has now been ended (verses 10 and 11). Through the atoning death of Christ any accusation against believers has been eliminated. In Colossians 2:14 and 15 we read: "... having wiped out the handwriting of requirements that was against us, which was contrary to us. And He has taken it out of the way, having nailed it to the cross. Having disarmed principalities and powers, He made a public spectacle of them, triumphing over them in it". In this way the believers could also overcome the devil through the blood of the Lamb and the word of their testimony (Rev 12:11).

The result was that the devil and his angels were caused to focus more sharply on the area in which they were still able to operate, namely amongst people on earth. He does this with great fury, knowing that his time is short (Rev 12:12).

3.4 The restraint or binding of Satan

Not only was Satan cast out of heaven, but God limited the power of the Evil One with a view to Christ's first coming and for the sake of his church, so that Satan cannot give free rein to his desires. For the sake of the chosen, God restrains the Evil One and limits his ruinous and sinful power so that everything is not destroyed. This evinces God's grace which is always at work.

We read how Satan could only do to Job what the Lord allowed him to do (Job 1:12). God has also curtailed the destructive force of the Evil One in nature. The disruption in creation and the corruption to which it has been subjected (Rom 8:19–22) can be linked to the devil's fall. In his providence God ensured that a reasonable measure of order and continuity of the natural processes remained intact. This makes life on earth still possible. It also enables man, by means of science and culture, to improve his sojourn on earth and make it more pleasant.

God guards his church and will not allow the *gates of hell* to overcome it (Matt 16:18). The power of the Antichrist already at work in the world is restrained for this reason. The devil will have free reign only at the end of this earthly dispensation. And when Satan is released, he will mobilise all humanity against God's saints, and then he will be destroyed. This happens at the second coming of Christ (2 Thess 2:7–8; Rev 20:7–10).

Satan's restraint by the victorious Christ comprises one reason why the believer can resist him and even compel him to flee (Jas 4:8; 1 Pet 5:8). For those in Christ, Satan is a conquered and bound enemy.

All of this bears witness to the kingdom of God which is revealed on earth. Therefore, this victory over evil and over the power of sin is only enjoyed by those in Christ, who share in the kingdom of God. After all, this is the basis on which God's promises to the believers rest (Ps 91; John 16:33; Rom 8:31–39; Eph 6:13–14; 1 John 2:13; 4:4; 5:4–5; Rev 12:11).

4 STRATEGY AND METHODS OF THE EVIL ONE

The strategy of the Evil One has always been to lead man into sin. When he succeeds in doing this, the unavoidable judgement as required by God's righteousness, must follow. In this way Satan intends to destroy God's handiwork. Along with this, he wanted to bring man, the one created in God's image and likeness, under God's rejection and judgement. In this way he tried to render it impossible for a body and bride for Christ to be taken from humanity.

We have already pointed out that the power of the Evil One has been limited by God. For this reason the devil is unable to force people to sin. Moreover, he cannot compel people to reject the grace of God offered through Christ.

We have to understand Satan's strategy and method within the above framework. With regard to the unbelievers, he does his utmost to prevent them from coming to faith and repentance. He would prefer to see them becoming more and more sinful and ungodly. With regard to the believers, he does everything possible to make them backslide from their faith. Failing this, he at least endeavours to make life so difficult for them that they can barely serve the Lord and comprise a force within his kingdom. It is in this regard that Paul renders us a grave exhortation in Ephesians 6:10–20. In this passage of Scripture he points out that we are engaged in a battle. Primarily this battle is not against people or visible things but against the evil powers in the air. (The qualification *in the air* apparently refers to the fact that the evil forces are invisible to man.) In addition, he admonishes us not to carry on this struggle in our own strength, but only in the strength which the Lord supplies. This constitutes spiritual warfare. 2 Corinthians 10:3–4 reads: "For though we walk in the flesh, we do not war according to the flesh. For the weapons of our warfare are not carnal but mighty in God for pulling down strongholds." This concerns a struggle in which man has to remain standing *against the wiles of the devil* and at the same time should endeavour to bring everything under the rule of Christ (Eph 6:11; 2 Cor 10:5).

The fact that Paul speaks of the wiles of the devil shows us that here we are involved with a dangerous deceiver.[15] Ephesians 4:14 states that this *cunning craftiness* is aimed at deceiving man.

The devil has been a liar and deceiver since the time of Adam and Eve (2 Cor 11:3). His strategy and method is to conform people's will and spirit to his own through deception. Thus he obtains control over them whereby he can enslave their entire lives.

The Evil One attacks with a total onslaught so as to enslave people completely. His onslaught, like fiery arrows, can strike a person from any direction (Eph 6:16). Every kind of human behaviour, whether thoughts, desires, impulses, words or deeds can be affected by sin, that is, may be influenced by evil. For as many sins as are committed, just so many strategies and methods of the devil are involved. Therefore it is impossible to identify all the methods and crafty plans used by the devil. We can only focus attention on a few. Knowing about them can help people to avoid the snares of the Evil One. In 2 Corinthians 2:10 and 11 Paul warns us against the deception of Satan, and adds that we can know the plans which Satan devises.

When we analyse the strategies and methods which the devil uses, we find that these are very flexible. They are designed to adapt to every person's desires, circumstances, idea of religion, way of thinking, problems, social structures, etc.

Let us consider a few examples.

4.1 Satan uses some of man's natural impulses to cause him to sin

Since the time of Adam and Eve, Satan has used that which appears attractive and pleasant to man to tempt him to sin (Gen 3:1–6). We notice how the

15 Grimm describes the Greek word *methodeia* (wile) as follows: cunning arts, deceit, craft, trickery. *Greek-English Lexicon NT*, pp 395, 396.

Evil One (in the form of a snake) confronted Eve with two things. To begin with he tells her that if she should eat of the fruit of the tree of knowledge of good and evil, she will not die as God had warned. He assures her of the contrary. Instead of dying, her eyes will be opened and she will be like God, knowing good and evil.

Secondly, and related to the former, he draws her attention to the desirability of the forbidden fruit. Genesis 3:6 tells us what followed: "So when the woman saw that the tree was good for food, that it was pleasant to the eyes, and a tree desirable to make one wise, she took of its fruit and ate. She also gave to her husband with her, and he ate."

We comment as follows on this: Striving to become increasingly like God is not necessarily wrong. In fact the Lord does want us to become more and more conformed to his Son (Rom 8:29). Neither is enjoyment of the good and the beautiful of life necessarily wrong (Ps 107:9). It is precisely according to God's will that we should enjoy everything with gratitude which He grants us or enables us to obtain. After all, He is the Giver of all good things (Jas 1:17). The problem, however, arises from the manner in which man wants to obtain the above-mentioned things. Does this conform to God's will or man's desires? Does man act in obedience to God's word or decide on his own what is good and beneficial? It is a case of either living devoted to and dependent upon God, or trying to put yourself on a par with God (as Satan himself did) and, thus, deciding for yourself what your life should be like.

To be conceited and proud is described in 1 Timothy 3:6–7 as the *snare of the devil.* Consequently self-improvement without God is the primeval sin which brought judgement and corruption upon mankind.

Ever since the devil has used this same strategy. We see this in the sins which people commit. Instead of following the way God prescribed for us, man chooses the direction dictated by his own desires and impulses.

The devil used this strategy when he wanted to tempt the Lord Jesus. We read that the Lord fasted for forty days and nights and became very hungry. The devil saw this as an opportunity to tempt the Lord Jesus with the following words: "If you are the Son of God, command that these stones become bread." (Matt 4:3). This would mean that the Lord should subject himself to the suggestions of the Evil One and thus avoid the suffering which his divine mission entailed. The Lord Jesus answered the Tempter by pointing to a higher principle than the mere satisfaction of natural desires. One has to live according to the word of God (Matt 4:4).

We also read how the devil tempted Judas by placing the suggestion to betray Christ into his heart, and thereby procuring money for himself. The devil knew that Judas was avaricious (1 Tim 6:10) and that his lust for money afforded the gateway to persuade this disciple to carry out his evil intent. In the same way the devil is still tempting people to enrich themselves contrary to the will and law of God. God's will is that man must not benefit himself at the cost of others, but that he should work for his money (Eph 4:28).

This same strategy is used by the devil with regard to sexual and other natural desires. To have sexual desires is not wrong; God made people in this way. However, the Lord has declared that these desires are to be satisfied within the marriage bond. Satan, however, tries to present another and more exciting way to satisfy this desire. Consequently people are tempted to fall into the evil trap of fornication and adultery (1 Cor 7:5). Paul sees this satanic temptation as particularly dangerous to young widows – and by implication to young widowers too (1 Tim 5:14–15).

Some people are so obsessed with the need to satisfy their natural desires, that these things become more important than God himself. In Phillipians 3:19 Paul speaks of those whose stomach has become their god. In Romans 16:18 he puts it thus: "For those who are such do not serve our Lord Jesus Christ, but their own belly, and smooth words and flattering speech deceive the hearts of the simple." When one chooses anything else above God, even the satisfaction of his natural impulses, one is doing precisely what the devil wants him to do.

The sinful ways in which natural desires are satisfied is romantisised in this world or presented as an accepted way of life. We see this in advertisements, pornography, so-called works of art, solicitation and prostitution, unethical presentations to make money, suggestions of how to manipulate people in your own interests, videos which portray immorality and violence, etc. All these things show us who the god of this world is (2 Cor 4:4).

One of the things which makes one susceptible to such temptations is what Paul calls *foolish talking* or *coarse jesting* (Eph 5:3–5).

These temptations, together with the concomitant dangers, are not limited to unbelievers only. The fact that the regenerated person still experiences a problem with his *old nature* gives the devil something which he can use to tempt every believer (Rom 7:19–26). Thus Paul also admonishes the believers to reckon their old sinful nature dead and crucified with Christ (Rom 6:4–7).

4.2 Satan exploits man's religious predisposition

When Satan deceived the first human couple in the Garden of Eden, he did not deny the existence of God. In the beginning Adam and Eve had personal contact with God so that they could never question his existence. After the tragic fall of man, man no longer enjoyed that direct fellowship with God. What man, however, did not lose was his religious predisposition. This is because man is spirit and is created in the image and likeness of God. For this reason man inherently seeks contact with a transcendental force and practises one or other form of religion. Consequently those who do not know the true God worship that which is no god (Gal 4:8).

Satan exploits this religious predisposition. The mere fact that man is naturally disposed to religion does not pose any threat to the Evil One. What he does find dangerous and intolerable is when people encounter the true God and worship Him. For this reason he tries to prevent man from experiencing God's grace through Jesus Christ. One of his strategies in realising this aim is to inculcate a false idea about God. When people do not practise a religion in spirit and in truth, they cannot be delivered from the Evil One in any case. We must never lose sight of the fact that Satan abuses man's religious inclination for his own ends.

4.2.1 The non-Christian religions

This evil strategy can be discerned in various kinds of religious behaviour. Firstly, we encounter it in religions where faith in Christ as the Son of God, the Saviour of the world, is denied.

In these religions we find some concept of God along with moral principles which, from a human perspective, are good. The hope of a blessed hereafter which these religions offer however, is based on own merit or good works. This is contrary to what the Bible teaches us about faith in Christ as the only way to salvation. Though wrong, these religions offer their adherents at least some direction in life and satisfy their religious predisposition. In addition they order their society and lend a certain meaning and value to their lives. By means of these religions the devil makes sure that man does not look for Christ as his only salvation (John 8:24). We believe that Christ is the only One who can redeem us from sin and reconcile us with God. Therefore, if the devil can keep Christ and his gospel out of religion, he has achieved his purpose.

The Bible teaches explicitly that mankind is divided into two groups: those who believe in the Son of God and those who do not believe in Him (John 3:18). One either has the Spirit of God or the spirit of Antichrist (1 John 4:1-6). The spirit of the Antichrist is the spirit of the Evil One, no matter in which refined form it may appear.

We find all kinds of occultic practices in some of these non-Christian religions. This pertinently stresses the demonic deception. (We shall deal with this aspect far more extensively later on.)

Although the teaching and religious practices of non-Christian religions cannot really bring about true salvation, we do not maintain that everyone who has not yet encountered the gospel of Christ will perish without ado. We do not know which personal revelation God has graciously granted to individuals outside of Christendom. It seems possible that people could have come to true faith in God or share in his grace even though they may have never heard the gospel. God is not bound to the proclamation of the gospel as we know it in order to save people. In this respect, we are reminded of the redeemed of the Old Testament period and of babies who die in infancy. These are indeed saved on the grounds of the atoning death of Christ, even though they were never taught this.

4.2.2 Adhering to non-Christian beliefs besides the Christian faith

The devil knows full well that if, besides Christ, you also serve idols and adhere to all kinds of non-Christian practices, you are unacceptable to God. An example of such syncretism is when people pretend that they are Christians, but at the same time believe in ancestral spirits who exercise a great influence on their lives.

It is with a view to any form of syncretism that Paul warns the Corinthians not to profess Christ on the one hand while on the other hand, have fellowship with idol worshippers. In 1 Corinthians 10:20 and 21 he says: "And I do not want you to have fellowship with demons. You cannot drink the cup of the Lord and the cup of demons; you cannot partake of the Lord's table and of the table of demons." In 2 Corinthians 6:14–16 he also says: "Do not be unequally yoked together with unbelievers. For what fellowship has righteousness with lawlessness? And what communion has light with darkness?" (Cf also Matt 6:24.)

The darkness described in the afore-mentioned Bible texts refers to the sphere of influence of the Evil One (Eph 6:12). For this reason sin is also called the works of darkness (Rom 13:12; Eph 5:11).

The devil can even appear in the form of an angel of light in order to seek fellowship with believers. In Acts 16:16–18 we read about a girl with a divining spirit who exclaimed about Paul and Silas: "These men are the servants of the Most High God, who proclaim to us the way of salvation." What she said was indeed the truth; the agent who made her say this, however, was the Evil One. Paul discerned this and therefore he was very alarmed about what she had done. He then cast the evil spirit out of her. Paul did this in spite of the serious repercussions that this would have for him and his fellow traveller.

4.2.3 Through the sins of ministers of the gospel

Satan also knows that Christendom can be paralysed if ministers of the gospel do not live pure lives. If the devil can use a believer, and in particular, a minister of the gospel, he will never let such an opportunity slip. We see how the devil even used a disciple of the Lord to make a wrong decision. Apparently in a moment of pride or wilfulness, or simply in a moment of imprudence, Peter reprimanded the Lord Jesus when he spoke of his own coming suffering and death. The Lord's reaction (Matt 16:23): "Get behind Me, Satan! You are an offense to Me, for you are not mindful of the things of God, but the things of men," makes it clear that here we are witnessing a case of demonic influence.

It is not only the irresponsible and inappropriate pronouncements of ministers of the gospel which serve the cause of the Evil One. Ministers also serve Satan's cause by acting in ways unworthy of their office and through immorality. What immense damage has Satan not already inflicted on the church of Christ through the sins of prominent preachers! The devastation is at its worst when it affects young believers who have trusted such a minister to be a man of God. In this regard, we quote the condemnatory words of the Lord in Matthew 18:6: "But whoever causes one of these little ones who believe in Me to sin, it would be better for him if a millstone were hung around his neck, and he were drowned in the depth of the sea."

Satan also uses religious people to demonstrate a false piety. In this way authentic piety is supplanted while a distorted image of genuine religion is presented. An example is when people try and pretend how spiritual they are by virtue of their external religious behaviour. At the same time these people may be bound by all kinds of sins. This was pointed out by the Lord Jesus with reference to the Pharisees and the scribes (Matt 23:23–33; Luke 11:37–54). This hypocrisy also reveals the Evil One's deception and deceit. If he can find *blind leaders* to mislead people (Matt 23:24), everyone will certainly perish!

4.2.4 Satan uses false brethren and strange teachings to lead people astray

In addition to what has already been said about Satan's deceit and onslaught in the area of religion, we must also mention false brethren and doctrines. The introduction of false elements is a strategy used by the Evil One to try and destroy the Church of Christ from within.

False brethren and strange teachings are only relevant where a fellowship of true brethren already exist and where the true gospel is proclaimed. Falsehood always feeds on truth. It pretends to be the true gospel, but in reality it is a totally different gospel (Gal 1:6–12). They confess piety, but they express something totally different in their lives (Tit 1:16).

In order to make the false element attractive and successful it has to resemble the genuine. The false also has to be mixed with the genuine. This state of affairs is illustrated by the parable of the weeds and the tares told by the Lord Jesus (Matt 13:24–30). The kingdom of Heaven, as it exists on earth, is compared to a cultivated field. The farmer sows good wheat on the land but when night comes and no-one suspects any evil, the enemy arrives and sows tares in the same field. The wheat and tares then grow together in the same field until the time of the harvest. Then comes the separation. The tares are gathered and burnt, while the wheat is stored in a barn (cf also Matt 13:47–50). The term *enemy* refers to the devil and the *tares* refer to the children of the Evil One (Matt 13:38).

This parable shows us how false brethren can co-exist with true believers in the same congregation. The devil smuggles them into the congregation secretly and unobtrusively. There they pretend to be prophets of God (2 Pet 2:1–18; Rev 2:14, 20). Once they have found a place in the congregation, they unobtrusively introduce false doctrine. They deny the Lord and despise his authority as it pleases them. They exploit the believers with their appearance of piety. In addition they are challenging and haughty. They like to carouse, are deceitful speculators, lustful and bring all kinds of shame upon the congregation (2 Pet 2:1–22; cf also 1 Cor 5:11; 2 Thess 3:6).

The Lord Jesus describes such people as wolves in sheep's clothing (Matt 7:15–23). Outwardly they appear to be true believers. They prophesy in the name of the Lord, they do miracles and even cast out devils, (Matt 7:22–23). In reality they are agents of the Evil One and will be condemned along with the evil angels (2 Pet 2:3–4). In the meanwhile however, they mislead the unwise and ignorant as well as the credulous and weak in the faith (Rom 16:18; Eph 4:13–14) with their friendly words (2 Pet 3:16).

Another characteristic of the false brethren is that they occupy themselves with all kinds of trivial matters which can only cause dissension (1 Tim 1:3–4).

Just like Satan, the false brethren also misuse Bible verses which they then interpret to serve their false ideas (Matt 4:6). For this reason Paul declares " that in latter times some will depart from the faith, giving heed to deceiving spirits and doctrines of demons". (1 Tim 4:1). James describes this as the wisdom which does not come from God, but which is *earthly, sensual and demonic.* On account of the envy and selfishness of their hearts, confusion reigns together with every evil thing (Jas 3:15–16).

A powerful weapon used by the devil to lead people astray is the miracles which he performs through his agents. The devil is a master at imitating the works of God. He does miraculous things whereby he easily misleads the ignorant and those without the ability to discern (Exod 8:19; Acts 8:9–10; Rev 16:14). Other things which occur supernaturally can also be mentioned. Here we think of the different occult practices such as divining

(Lev 19:31), the summoning of spirits (Isa 8:19) and so on. We read that when the Antichrist and the false prophet appear they will entice many people to follow them by means of miraculous signs which the devil will enable them to perform (2 Thess 2:9; Rev 16:14).

In addition the devil does not only deceive people through false brethren and evil doctrines. He also makes use of atheistic philosophies and so-called scientific conclusions to destroy faith in God and the Bible (Col 2:8). This is then another method with which the devil blinds the senses and the minds of unbelievers so that they cannot see the light of the gospel (2 Cor 4:4).

The notion that God will always make you glad whereas the devil will always make you sad, can be very misleading. There can be "blessings" like miraculous healings, baptisms in the Holy Spirit, joy and excitement in worship, et cetera which do not originate from a true working of the Holy Spirit. Such blessings may be more the result of something which a preacher – mostly unwittingly – induces upon a congregation through psychological manipulations. Those eager to have miracles at all costs, and who are gullible enough to ascribe any extraordinary happening to the manifestation of the Holy Spirit, will easily become victims to feigned wonders. When they ultimately discover that things did not work out as they have thought, disappointment and disillusionment follow. In this manner the Evil One can cause great spiritual damage to the church.

Regarding those who were used by the devil to mislead others, while becoming rebellious, Paul instructs that they be admonished in a spirit of *gentleness*: "In humility correcting those who are in opposition, if God perhaps will grant them repentance, so that they may know the truth, and that they may come to their senses and escape the snare of the devil, having been taken captive by him to do his will." (2 Tim 2:25–26). It may also be necessary to reprimand somebody like this very severely (Tit 1:13; 2:15). Sometimes the correct behaviour is to withdraw from such a person (2 John 10) and deliver him to Satan, with the hope that he will come to repentance on account of his suffering (1 Cor 5:5).

4.3 Satan uses the sinful world to try and destroy faith in God

In John 12:31; 16:11 the Lord Jesus refers to the devil as the prince of this world. This does not mean that the visible world – the earth and everything on it – belongs to the Evil One. God is still the Lord and Sustainer of everything, and ultimately his will shall triumph over everything. It is only with reference to the sinful state in which the world finds itself that the devil is called the prince of this world.

It is in relation to this then that we have to understand the strategy and the method of the Evil One. His evil influence manifests itself in different ways in this sin-ravaged world.

4.3.1 Onslaughts endured by saints in this world

The story of Job shows how Satan attempted to destroy the faith which good people have in God. In a dialogue with the devil the Lord pointed out the virtues of Job – his piety, sincerity and service to God. In addition he avoided all evil. In response Satan alleged that Job was only doing all this because the Lord had protected and prospered him. If this grace was removed from him, he would curse God in his face. Thereafter we see how the Lord allowed Satan to rob Job of his children and possessions (Job 1:6–19). In these events we notice how the Evil One uses wicked people (robbers). He could even harness nature (lightning and tornado) to harm this pious man.

We read further that Job did not reproach God in spite of all these adversities. On a later occasion when God pointed out these facts to Satan, he was immediately ready to launch a further onslaught on Job. This time he said to the Lord that if Job's body was affected, he would most certainly curse God. Once again the Lord permitted him to employ his evil strategies. Satan afflicted this pious man with painful sores from head to toe. It was so bad that Job went and sat on an ash-heap and scraped his body with a potsherd (Job 2:8). Even when afflicted so severely Job did not turn against God. We only read that he wished he had never been born (Job 2:1–8; 3:1).

Furthermore we see how the one who had to support him, namely his wife, also turned against him. In Job 2:9–19 we read: "Then his wife said to him, 'Do you still hold fast to your integrity? Curse God and die!' But he said to her, 'You speak as one of the foolish women speaks, shall we indeed accept good fromGod and shall we not accept adversity?' In all this Job did not sin with his lips."

Even today the Lord still allows Satan to launch these evil attacks on believers. Blessed are those who remain faithful to God in the midst of such afflictions!

We also read how the apostle Paul was attacked by an angel of Satan. What this angel did to him is not described. This affliction could have taken various forms. It was possibly of a physical nature or may have involved hostile people or maybe just trying circumstances under which Paul had to work. It was, however, of such a serious nature that it was compared to the ongoing suffering of a thorn in the flesh (2 Cor 12:7–10)

However, all these onslaughts could only take place within the boundaries set by God. Furthermore we should note that Satan's onslaughts upon these believers had the opposite effect to that intended. By the grace of God any onslaught upon believers can turn out positively. The faith of these people can be purified and tested by these afflictions so that they emerge as conquerors from the struggle (1 Pet 1:6–7).

4.3.2 Because of his power in this sinful world, Satan sets all kinds of stumbling blocks in the way of believers

Besides the direct onslaughts upon people, Satan also lays stumbling blocks in the way of those who serve the Lord. We refer to a few examples from Scripture.

Paul and his companions eagerly wanted to visit the congregation in Thessalonica. However, they could not manage this. The reason for this is given in 1 Thessalonians 2:18: "Satan hindered us."

Paul also experienced resistance when he proclaimed the gospel to Sergius Paulus. We read how a magician by the name of Elymas tried to prevent the governor from believing in Jesus. Thereupon we see how the apostle rebukes him with the words: "... full of all deceit and all fraud, you son of the devil, you enemy of all righteousness ..." (Acts 13:10). After this he struck the magician with blindness.

We have to accept that everything that restricts the gospel is due to the Evil One. By this we mean things like wars and uprisings, ungodly governments, oppressive laws which make it difficult for the church to fulfil its calling, state interference in religion, the legalising of practices such as gambling, prostitution or anything else contrary to Christian ethical principles.

4.3.3 Satan deceives mankind by the spirit of this world

One of the most powerful means by which the Evil One keeps people from a personal encounter with the Lord and a life in the Spirit is the influence which this world exerts upon them. Mankind with its diverse peoples, societies, nations, cultures, et cetera forms a corporate unity and a power. As a person can exert an influence on those with whom he is dealing, so this world also comprises a kind of collective and overarching personality. It reveals a particular spirit, ethos and attitude which exerts an influence on people. By word and deed this sinful world proclaims an anti-Christian and a God-dishonouring way of life.

Judged in the light of Holy Scripture, the spirit of the world is pitted against the Spirit of God. Paul speaks of this in 1 Corinthians 2:12: "Now we have received, not the spirit of the world, but the Spirit who is from God ..." In this connection there is mention of *the present evil world* (Gal 1:14). We are reminded that we cannot love God and the world (1 John 2:15), and he who wants to be a friend of the world is an enemy of God (Jas 4:4). Furthermore we are exhorted to "... keep oneself unspotted from the world." (Jas 1:27). The spirit of the world is that of the prince of this world, namely the spirit of the devil.

4.3.3.1 The world as spirit and the world as a location

In this connection we have to distinguish clearly between the world as spirit and the world as a place. Because the general spirit of the world is anti-Christian, it does not imply that the world is therefore totally evil and worthless. There are many things in the world which are good, beautiful and true. Through God's control and providence in this very sinful world there is still a measure of order, peace, progress, care, pleasant experiences, opportunities to live good and fruitful lives, and many other things to be grateful for. This grace and providence of God offers the necessary opportunity for the continuance of the true Church until the return of Christ. This same grace and providence also, however, provides room for the Evil One to fulfill his role. True, God still tolerates the evil forces, but this is only until the return of Christ, whereafter these will be destroyed.

The spirit of the world is evil not only because the greater majority of people are sinful, but also because this majority rejects Christ. Thereby they reject God and yield themselves to the devil. This brings the terrible judgements of God upon the earth. We read about this in the Book of Revelation.

Even though humanity still enjoys so much favour, grace and goodness from the hand of God, the majority of people do not give the Creator and Sustainer of all things the honour due to Him (Rom 1:18–21). All honour goes to man with his science, ideologies, different systems, technological ability, artistic sense, planning and so on.

4.3.3.2 The influence of the spirit of the world on Christianity

Since the world does not know God, it neither knows nor understands the true fellowship of believers (1 John 3:1). Because true believers do not want to bow before the idols of this world, the world hates them (1 John 3:13). (By idols we mean all things man substitutes for God.) In the eyes of the world, true believers are a spectacle and like scum (1 Cor 4:9, 13.)

However, it should be clearly said that the greatest danger which the spirit of the world poses for the Church of Christ does not lie in violent behaviour. The actual danger is that the world tries to gradually condition believers to think, feel, desire and act, like itself. Yes, it wants to transplant its spirit into all people. This spirit is like a tremendous vortex which wants to suck everyone into its influence and control. This onslaught on the spirit or heart of man usually happens in a gradual, subtle and obscure manner. As one is exposed to the spirit of this world, one is inclined to assume the direction and way of life of this world.

Experience dictates that one is commended and rewarded by the world in proportion to the measure in which one identifies with the spirit of the world and participates in the world's God-dishonouring practices. He who does not conform to the spirit of the world is dismissed as rebellious, narrow-minded, eccentric, maladjusted or simply irrelevant.

Even prominent Christians have succumbed to the pressure of the spirit of the world. We read in Galatians 2:11–16 how the apostles, Peter and Barnabas, behaved hypocritically on account of social pressure. At first they used to eat with believers converted from heathendom. However, when Jewish Christians appeared, they withdrew from those Gentile Christians out of fear of going against Jewish tradition.

The pressure of the spirit of the world has caused numerous people to exchange the narrow and arduous path of the gospel for the broad and easy way of the world (Matt 7:13–14). For this reason people like Demas turned their backs on Christianity and once again identified with the sinful world (2 Tim 4:10).

Conformity to the world inevitably leads to regression in faith. For this reason we have to heed Paul's words in Romans 12:2 seriously: "And do not be conformed to this world, but be transformed by the renewing of your mind, that you may prove what is the good and acceptable and perfect will of God."

4.3.3.3 Comparison between the Spirit of God and spirit of the world

In order to understand the spirit of the world correctly, we have to understand it in contrast to the Spirit of God who is at work in people. The Spirit of God focuses one's interest and dedication on things of eternal value. Colossians 3:2 declares: "Set your mind on things above, not on things on the earth." (Cf also Rom 8:5–9; Phil 3:19.) He who lives through the Spirit of God will do everything here on earth according to the great commandment, namely to love God above all else, and to love your neighbour as yourself (Matt 22:36–40).

By contrast, the spirit of the world works conversely. It calls for your attention and abilities to be focused on the temporary things of this world. It desires that you be so preoccupied with making money, with sport, technological achievements, scientific knowledge, art, entertainment, politics or whatever human endeavour there might be that God and his gospel will be on the periphery of your life or even completely eliminated from it.

The activities referred to in fact have a place in our earthly existence. It is only when they become so important next to or in place of the gospel truths that they are nothing less than idols. Then they become part of the vortex of this world whereby man is sucked away from God.

However, this also happens where the Christian doctrine and religion is abused in order to further one's selfish aspirations or to serve a political and social aspiration. The correct practise of politics, business, social issues, et cetera occurs only when these things take place within the framework of the gospel and in the service of eternal values.

The Spirit of God leads a person to the knowledge of God and usefulness in the kingdom. Where the spirit of the world still does make room for religion, it is limited to certain occasions. Together with this, religion is not allowed to interfere with everyday activities or judge them.

The Spirit of God convinces man that he is sinful, under the power of the Evil One and subject to condemnation. By this same Spirit one is reborn and becomes a recipient of the new life through Christ. To the extent that one lives for Christ and not for oneself only, one becomes an authentic and complete human being. In this way one obtains genuine fulfilment in life.

The spirit of the world works in the opposite direction to that portrayed above. It does not seek after God, the Creator and the Sustainer of all things, but it seeks after the god within you, that is, after your own abilities.[16] Through your own philosophy of life and through everything you do, you have to promote your own image and esteem in the eyes of others and yourself. Serving yourself is your first priority. If you treat others well, it is to improve your own self-image and acquire certain merits. This attitude does not care about sin, an evil force which keeps you under its power and impending judgement. Any weakness or incompleteness which one may experience is only a stage in an evolutionary developmental process.

All the afore-mentioned ideas can be summarised in a few words. According to the Word of God, the Lord alone is your Saviour. By believing only in Him and dedicating your life to Him are you delivered of evil things and do you receive fulfilment in life. In contrast, the spirit of the world declares that you are your own saviour and have the right to choose without regard to God's law what is good, pleasant and beneficial. In addition you should strive to be as independent as possible of forces above and around you. The self-made man is the ideal man! A person with this philosophy of life is so preoccupied and blinded in his mind that he cannot perceive the things of God. His heart is hardened so that he has no desire for a whole-hearted conversion. Such a person is spiritually dead (Eph 2:1; 1 Tim 5:6). This is the condition which the devil wants to create in every person.

4.3.4 The strategy of Satan is to lead from evil to more evil

The devil is never satisfied with a single evil action. The prince of this world desires that one evil should yet lead to another. In this way he can easily

16 This is the typical base of the New Age movement.

achieve his objective with man. We refer to two examples of how this strategy of the Evil One works.

4.3.4.1 By requiting evil for evil

One way in which the devil achieves his purpose is by inducing people to repay evil with evil. This can take many forms. We refer to a few.

First and foremost it implies paying people back in their own coin. When somebody slanders you, damages or robs your property, assaults you or does any other evil to you, then you should retaliate.

This seems logical and fair. However, the fact is that such an evil action leads in turn to further evil actions. And so it becomes a self-perpetuating force until the maximum possible evil has been done. This is precisely what the devil wants.

But what should a true Christian do under such circumstances? Obviously everything should be done to bring an end to an evil deed. In addition the initiator should be identified and punished. This punishment is, however, the task of the authorities and eventually of God himself. This does not imply that you should be passive under the onslaught of evil. On the contrary, one should combat all forms of evil, also those aimed at oneself. You should do this within the prescriptions of the Bible and in accordance with the stipulations of the laws of a country. In doing so one should, however, beware that evil does not take root in one's own heart. In combating the Evil One you should not become evil in your own heart or spirit.

One should also remember that the Evil One, by means of external circumstances and the injustice done to you, will try and incite you to relinquish your position of faith. Once your spiritual ramparts are down, you are open to further assault from the Evil One upon your life. Then he gets the opportunity to follow the first blow which you received with a second – a knock-out blow aimed at your heart. It is with regard to that that we have to heed the words of Ephesians 4:26 and 27: " 'Be angry, and do not sin', do not let the sun go down on your wrath, nor give place to the devil."

Among the appropriate measures to combat injustice, there are also some of special importance. The Bible teaches us not to harbour any bitterness in our hearts, but to forgive. You should even be willing to ignore injustice or endure it if by doing so you can strengthen your spiritual condition and testimony (1 Cor 6:7; 1 Pet 2:19). By repaying evil done to you with good, you can break through the evil cycle spoken of above. This is in fact what Paul advocates in Romans 12:19–21: "Beloved, do not avenge yourselves, but rather give place to wrath; for it is written, 'Vengeance is mine, I will repay,' says the Lord. Therefore 'If your enemy is hungry, feed him; If he is thirsty, give him a drink; For in so doing you will heap coals of fire on his head.' "

4.3.4.2 By way of a wrong reaction which makes you more defenceless against the onslaught of the Evil One

In this world there are so many evildoers, thieves, perpetrators of violence, insurrections and all other kinds of evil things that people feel threatened and insecure. It may even cause the believers' hearts to be filled with fear. In 2 Timothy 1:7 there is reference to a *spirit of timidity* – something that is not of God. When Christians react in fear, it implies that their faith and their spiritual lives are collapsing. It implies that they are finding it difficult to believe in the love and the care of God. In 1 John 4:18 the apostle contrasts fear with love. These two cannot co-exist; it is either the one or the other (cf also Job 3:25; Rom 8:15).

In this respect we must also refer to periodic persecution of the church and its members. Part of the strategy of the Evil One is to allow a spirit of confusion, doubt, despair and fatalism to take root among Christians through persecution. The Tempter wants Christians to be prepared to relinquish their faith for the sake of security and acceptance by the world. Even though they would not deny their faith, the Evil One at least wants to see persecuted Christians reduce their activities to a minimum.

Paul encourages the Ephesians as follows: "Therefore I ask that you do not lose heart at my tribulations for you, which is your glory" (Eph 3:13). To the Thessalonians who are being persecuted, he writes: "For this reason, when I could no longer endure it, I sent to know your faith, lest by some means the tempter had tempted you, and our labor might be in vain." (1 Thess 3:5). Romans 8:35 makes it clear that it is the purpose of the Evil One to separate Christ and his disciples by means of persecution.

As the end of this dispensation approaches, the world will become more and more evil (Rev 22:11) and the onslaught against the true believer will become more severe. The Lord, however, gives us the assurance that this can only take place within certain parameters. In 1 Corinthians 10:13 we read: "No temptation has overtaken you except such as is common to man; but God is faithful, who will not allow you to be tempted beyond what you are able, but with the temptation will also make the way of escape, that you may be able to bear it."

CHAPTER 7

Demonology: Demonising

Thus far we have dealt with the more indirect way in which man encounters the Evil One. However, the devil and his demons can also be revealed to man in a more direct and more personal way. This takes place when the devil, via his demons, works in such a way in a person's life that this evil force can be discerned more readily. This is known as *demonising,* derived from the Greek word *daimonizomai.*

We are talking primarily about demonising in the general sense. By this we mean that the Evil One controls one to such an extent that one in fact becomes demonic. This usually takes place when a person identifies himself with the Evil One or with evil things with the aim of organising his life accordingly.

Secondly, we speak of demonising in a more specific sense. This takes place when the Evil One's grasp on the person's life results in that person showing psychological deviations and at times insanity. In this case we specifically refer to the pathological condition caused by demonising. In order to distinguish this from demonising in general, we refer to it as *demonomania.* The word implies that a person is psychologically ill because of the presence of demons in his life. Bible translations talk of *demon possession.*

1 A GENERAL EXPLANATION OF DEMONISING

People are considered to be demonised when they have gone so far in the way of sin that they are identified with evil itself. It is, in other words, a very serious state of sin. Not only does such a person sin like any other sinner, but sin is so prominent in his life that the Evil One is recognised in him. This form of demonising can take on different forms.

1.1 When one's behaviour becomes demonic

We mention the case of Judas Iscariot as an example of this. In John 13:2 we read: "... the devil having already put it into the heart of Judas Iscariot, Simon's son, to betray Him". It was not a case where the devil forced him to act in this way. It was initially a temptation. However, it did not remain that way; Judas accepted this evil suggestion. He made this inimicable attitude toward the Lord Jesus and the intention to betray him part of his life's plan. When the Lord Jesus declared (John 13:21) that one of his disciples would betray him, Judas was thoroughly aware that the Lord was referring to him. He apparently did not heed this since he was already firmly resolved to carry out his evil plan. Then the Lord Jesus specifically identified Judas as his betrayer by giving him a piece of bread. Judas took the bread and with that openly acknowledged what was in his heart. In verse 27 we read: " Now after the piece of bread, Satan entered him." The Lord Jesus already knew beforehand what was going on in the heart of Judas. For this reason He referred to Judas as *a devil* (John 6:70).

This kind of diabolic behaviour directed against God and Christendom can be perceived with regard to many. We find this diabolic spirit among the Pharisees and chief priests of the Jews. The Lord still had mercy upon the heathen and unbelievers and brought them the gospel of salvation. The Pharisees and Sadducees and their followers, He refers to as a *brood of vipers* (Matt 3:7; 12:34), whilst He addresses them as follows (John 8:44): "You are of your father the devil, and the desires of your father you want to do."

This type of identification with the devil will find its acme in the Antichrist and the false prophet (2 Thess 2:3; 2 John 7; Rev 19:20). Whilst the Antichrist is yet to come, many antichrists have already appeared since biblical times (1 John 2:18). They have the spirit of the Antichrist – which is actually the spirit of the devil or the dragon (Rev 13; 1 John 4:2–6). We notice this in certain Roman emperors and other bygone rulers who cruelly persecuted Christians; dictators who came later and caused indescribable misery; atheistic philosophers and scientists who vehemently attacked Christendom in their teachings; evil church leaders who persecuted and tortured believers and various other enemies of the true Church.

The demonic in human behaviour also comes to the fore where the spirit of the mob is incited by hatred, anger and revenge. Under these circumstances people who would otherwise still act in a reasonably civil way can behave like animals, totally uncontrolled and without any sense of responsibility. In this raging frenzy destruction, death and misery can ensue which can only be described as demonic.

Something similar is also evidenced in the insanity of those enslaved to alcohol and drugs. This also happens where jealousy, selfishness and other immoral deeds begin to obliterate the limits of normal behaviour.

1.2 Idolatry

In 1 Corinthians 10:19 and 20 we read: "What am I saying then? That an idol is anything, or what is offered to idols is anything? Rather, that the things which the Gentiles sacrifice they sacrifice to demons and not to God, and I do not want you to have fellowship with demons." We find that people in biblical times worshipped Baal, Moloch, Dagon, Mammon, the sun and many other replicas and objects as gods. They also made images of these gods and bowed down before them. Together with this idolatry the Evil One was revealed. This was not only found in the false nature of their religion, but also in the insanity which gripped the worshippers. All kinds of excesses occurred in this immoral situation.

Idolatry of a similar nature is still found today among certain communities. For instance, we think of the totems erected for the spirits, the worship of ancestral spirits and various forms of animism.[1]

Idolatry, however, is not limited to the rituals of primitive societies, but also occurs amongst so-called civilised people. This happens when one devotes one's life to anti-god and sinful things and when one finds satisfaction in them. In Ephesians 5:5 we read: "For this you know, that no fornicator, unclean person, nor covetous man, who is an idolater, has any inheritance in the kingdom of Christ and God." The fact that the desire for these things is called idolatry (covetousness) indicates that any desire for sinful things which later control your life is actually idolatry. Through these things people become slaves of the demonic. Indeed one can describe such enslavement in psychological, physiological and other scientific terms. These proportions assumed by sin in a person's life are visible signs of the devil's control.

We mention the following as further examples of this kind of idolatry or servitude to evil: the irresistible desire to commit immoral deeds, use drugs, exploit your neighbour for your own benefit, yield to any sinful or destructive impulse, make money through illegal and socially damaging ways and so on. We cannot in fact equate each of these irresistible urges with a specific demon which has taken up its abode in a person's life. However, it is quite appropriate to speak of a demon of money, a demon of immorality, a demon of drink, a demon of power etc. All these things speak of nothing less than a sinful force – a manifestation of the Evil One himself in a person's life. Whether one believes in the devil or not, all these things come over one when one identifies oneself with evil things and gives them top priority. One may even go so far as to say that the worship of idols such as found in primitive societies is in no way more destructive and more evil than the idolatry practised in some so-called civilized circles.

1.3 Occultism

The word occultism is derived from the Latin *occulo* which means: to cover something. Occultism is used for practices to employ powers which are not accessible by natural or scientific means. These powers are so covered that they cannot be detected or understood by ordinary observation or scientific principles. It is something which lies in the realm of the so-called para-psychology. Although it is not presented as a kind of religion, nevertheless it is involved with mysterious psychic activities. Those who practise occultism maintain that they do not make use of supernatural things, but only exploit potential forces in nature and the human spirit.

Signs of occultism are also found in different religions, even in several which operate under the name *Christian.* An example of this is the faith in hidden powers which are supposed to reside in so-called holy things (the stone which would have fallen from heaven, certain trees or animals, relics or pieces of clothing of saints etc.) and powers connected to holy places (e.g. where Mary, the mother of Christ, would have appeared to people, where a holy one would have lived or been buried, etc.).

We can also mention many things under this topic such as: necromancy,

1 Cf. John L. Nevious, Demon Experiences in many lands, pp 7–128. Kurt Koch, *Between Christ and Satan*, 1961, pp 127–129.

alchemy, magic, astrology, telekienesis, clairvoyance, horoscopes and so on. We focus on a few general occultic practices and point out how these are instruments in the hands of the Evil One.

1.3.1 Necromancy or spiritism

Spiritists (they refer to themselves as spiritualists) pretend that they can come into contact and communicate with the deceased through mediums. Such a medium serves as a link between someone who specifically inquired about a deceased person and the deceased.

We discern the demonic in spiritism because of various factors. Firstly such a medium enters a state of trance where someone else speaks with another voice through him or her. (This would then supposedly be the voice of the deceased.) This phenomenon also occurs among demoniacs.[2]

Furthermore there are numerous testimonies where people who attended spiritistic seances developed a state of extreme anxiety. Where Christians involved themselves in these things their faith in God, the Bible and the church began to disappear.[3]

Subsequently we must point out that necromancers (spiritistic mediums) as well as those who solicit their help are severely condemned by God. (Lev 19:31; 20:6; Deut 18:10–12). This is a clear indication that this has to do with something evil.

1.3.2 Magic or sorcery

Magic is the practice whereby people are supposed to obtain power over mysterious forces.

1.3.2.1 Magical powers

Those with magical powers obtain these by means of cultic acts; entering a trance-like state; repetitive incantations; the use of magic formulae; the summoning of spiritual forces; the wearing of amulets; using spirit imbued objects (fetishes), potions prepared from all kinds of things (even human parts charged with a certain power) and so on. (We are obviously not referring here to the conjuring tricks of entertainment artists.)

By magical powers, one is supposed to ascertain where certain objects or people are (e.g. by means of casting bones), the cause of illnesses and plagues and how these could be avoided, the healing of certain ailments, the identification of hostile people or criminals, the ability to bring disaster to others, causing fertility among people and animals and so on.

1.3.2.2 Magic as a very old cult

Magic or sorcery is an ancient practice which already occurred among the nations in Old Testament times. For instance, we read of the Egyptian sorcerers who imitated some of the miracles performed by Moses (Exod 7:7, 22; 8:7). It is also mentioned that sorcery and divination were used as a weapon against Israel (Neh 23:23; 24:1). Furthermore we read about the Chaldean wise men, magicians and sorcerers who were reputed to have

2 Cf W C van Dam, op cit, p 74.
3 Cf Merrill F Unger, Demons in the world today, 1974, pp 194, 195. Kurt Koch, *Christian counseling and occultism*, 1965, pp 28–41.

been able to solve secret riddles (Dan 1:20; 2:1–27). In the New Testament we read of sorcerers who perplexed people with their extraordinary deeds (Acts 8:9–10; 13:6, 8).

1.3.2.3 White and black magic

In connection with sorcery we can distinguish between white and black magic. White magic operates under a religious cloak. Such magicians will even mention God and Christ. They will even quote certain texts from the Bible and other religious books. In particular, we here have witch doctors in mind who are supposed to be able to heal people.

Black magic has a deliberate connection with the devil. Demons are invoked in order to bring evil upon people and animals. This activity can also cause strange phenomena and peculiar noises, creating fear in people. It also makes use of evil powers in order to heal people.

Whether white or black magic, both operate according to occultic principles. Instructions appear in various writings, such as the Sixth and Seventh Books of Moses.[4] We read in Acts 9:19 how converted pagans burnt their books about magic.

A general practice in connection with magic is the ability to cast a magic spell over people or to free them from such a spell.[5]

1.3.2.4 Explanations for magical phenomena

Different explanations for these magical phenomena are given. They are regarded as something pseudo-scientific; the employment of hidden powers in man's spirit or subconscious; gifts given to people by God; the deeds of spiritual deviants; demonic operations; mere conjuring and the use of suggestion.

The notion that magic only appears amongst primitive people and is limited to pre-scientific times is simply not true. Today we are experiencing the revival of neo-paganism. Beside the worship of the medieval goddess *earth*, different forms of magic are also practised together with contemporary witchcraft and witches.[6]

It appears that people are fascinated by the mystical nature of occultism. Others again seek solutions to their problems and the healing of ailments in

4 Regarding this, F Unger writes: "This wicked treatise is falsely ascribed to Moses and attributes his God-given powers to magic (cf Exod 7:8, 9; 9:23). This deceptive work seeks to elevate Moses as the patron saint of all magicians." Op cit, p 92.

5 About casting and breaking of spells, Unger maintains: "A spell is produced by the release of demonic power through hypnosis, magnetism, mesmerism, or some other form of magic resulting in an extra-sensory influence. Conjurers, charmers, and others who dabble in both white and black magic frequently know how to cast and break spells. They can paralyze a person on the spot, cause a thief to be frozen in his tracks. A Swiss mesmerizer can cast his patients in a spell and force them to return to pay his fee if they forgot to do so.

Although both black and white magic use numerous other enchantments, yet the very heart of both branches centers in casting and releasing the spell. A spell can cause temporary blindness, deafness, dumbness, torpor, sickness, pain, etc. The symptoms will disappear when the spell is broken." Op cit, pp 92, 93.

6 Hobart E Freeman declares: "Incredible as it may seem, nevertheless even in this modern scientific age millions are now, or have been at some time, involved in some manner with ancient magic practices and rites, ranging from the charming of warts and burns to the use of spells, magic herbs, and hex signs on houses and barns (which can still be seen in some areas)." Quoted by F Unger, op cit, p 78.

magic. In addition, man is also inclined to search for a supra-rational truth behind the visible and material. When Christendom neither knows the mighty operation of the Holy Spirit nor allows it in their midst, the devil proffers his own counterfeit miraculous powers.

At times, particularly in Western countries, one would hardly even have heard of magic. This was certainly due to the prevailing rationalism and scientific thinking. Consequently people rejected the existence of occultic powers. For this reason there was no involvement with these things. Many things, however, indicate that an opposite tendency is increasingly coming to the fore. In contemporary times interest in the supra-rational and extrasensory influences are on the increase.

We can accept that many occultic phenomena, such as magic, may be relegated to conjuring, deception and all kinds of psychological factors, such as suggestion. However, there are also elements which suggest something supernatural which cannot be summarily dismissed. There are simply too many testimonies of reliable and unprejudiced observers which testify of similar demonic phenomena and events.[7] Actually the main issue is not whether natural or supernatural factors play a role in magic. What concerns us is how the Evil One, the Arch-deceiver, uses these things. He knows how to bring people under his power by means of magic.

1.3.2.5 Magic is strongly condemned by the Bible

Like idolatry, so also divination, conjuring, the explanation of omens and sorcery are strongly condemned by the Bible (Deut 18:10; 2 Chr 33:6; Mic 5:11; Rev 21:8).

That occultism must be linked to the Evil One is also clear from the fact that the power of sorcery can be broken by prayer and acts of Spirit-filled Christians.

It stands to reason that one would crave some knowledge of the supra-rational and spirit world. However, this knowledge cannot be obtained by empirically-scientific methods. It falls outside of the sphere of ordinary science. We can only take cognisance of these phenomena and understand their operation when we study them in the light of God's Spirit and Word. He, who desires to obtain this knowledge in another way, finds himself on a precarious way – the way presented by the Evil One.

In fact there are instances where it is alleged that people have been healed by means of magic and a solution was obtained for certain problems. This, however, does not change the fact that magic is an evil practice. Any advantage which people may gain from occultic practices such as magic is always at the cost of their own soul.[8]

It is also a fact that Christians who experience a relapse in their faith easily fall prey to occultism. In 1 Timothy 4:1 Paul speaks of those who will

7 Pastor J R Gschwend, a highly respected missionary, relates that he himself experienced how stones came up from the foot of a mountain and fell on a kraal (group of huts). One stone came through the wall of the hut in which he was, without leaving a hole in the wall. Only after they had rebuked the evil power in the Name of the Lord, did this assault stop. This miraculous deliverance resulted in the local chief and his people turning to Christ as their Saviour. *Light rays in a dark world*, pp 41–43.

8 Cf Kurt Koch, *Between Christ and Satan*, pp 84, 85.

become apostate from the faith in the last days and who will follow deceiving spirits and doctrines of demons.[9]

It is alleged that occultic abilities are carried over from one generation to the next, and that these powers may even be transmitted by means of one person's laying hands on another.[10]

1.3.3 Satanism

Satanism is an occultic movement in which people bind themselves to Satan and are then controlled by him. He is their god and they worship him. They also bring him sacrifices and execute his commands as communicated through satanic leaders.

1.3.3.1 A general survey

Occultism which is directly connected to Satan in one way or another has existed from time immemorial. It is, however, only since the second half of the 20 th century that satanism appeared in its present form. Since then it has spread all over the world, yes, even in the Western world. There was a time when political and ecclesiastical authorities doubted the existence of satanism. As a result of this they did not pay any attention to these phenomena. In addition, we should mention that satanists, in the early stages, behaved so clandestinely and covertly that people did not easily become aware of their existence. It was especially when ex-satanists (graciously delivered by Christ) began to tell what was really taking place that the reality of the existence of this evil practice actually reached the public ear.

It was Aleister Crowley (1875–1947) who to a large extent laid the foundation for present-day satanism. It was developed even further by Anton Szandor La Vrey (born 1930). He is also the author of *The Satanic Bible*. Under La Vrey's leadership satanistic cells and churches were established.[11]

Schisms in satanic leadership also occurred. So it happened that a certain Michael Aquino broke away from La Vrey and began the Temple of Seth. He maintained that the only true god was the Egyptian divinity, Seth, who has been accused of being Satan for 2 000 years by Christians.[12]

Satanism is also a hedonistic movement where people learn to succumb to all kinds of lust. In particular, we think of sexual perversions, the unbridled desire for power and the manipulation of people. All these things which are contrary to the Christian faith and life are applauded by satanism. The most general symbol of this evil practice is the encircled inverted pentagram. Sometimes a goat's head appears in the middle of the pentagram. There are also other signs and symbols such as the alteration or distortion of the Christian cross, with the specific aim of mocking it.[13] Satanism declares that Christ and his church holds no hope for man. As against this

9 W C van Dam remarks: "Also all the church fathers were convinced that every form of occultism, in as much it is not mere deception of men, were caused by demons. Even Augustine devoted a whole writing to it: About the soothsaying of demons *(De divinatione daemonum)*. Would the power of Satan not be revealed with all power, signs and lying wonders'? (2 Thess 2:9; cf. also Matt 24; 2 and Rev 13:13, 15)." (translation – F.P.M.) Op cit, p 80.

10 Cf Merrill F Unger, op cit, pp 82, 83.

11 Cf Keith Edward Tolbert, Satanism in America, *Charisma*, October 1992, pp 33–34.

12 For more information about satanic organisations we refer to John and Helen Gardiner, *Satanism: The Seduction of South Africa's Youth*, 1990, Struik House, Cape Town, pp 47–58.

13 Cf John and Helen Gardiner, op cit, pp 93–97. Keith Edward Tolbert, op cit, p 35.

they believe that if they promote the kingdom of Satan, they will be rewarded.

Tolbert points out that there are four forms of satanism in the USA. Firstly there are the organised satanist churches; secondly there are the solitary satanists (loner satanism), thirdly there are those intent on maltreating young children in rituals (satanic ritual abuse) and fourthly, the youth-culture satanism which propagates a satanic lifestyle among young people.[14]

1.3.3.2 How members are recruited and changed into satanists

Satan is the great deceiver and the father of all lies. Consequently he brings people under his influence by his great cunning.

Usually the journey to satanism begins when people involve themselves in ostensibly innocent matters, but which are in reality occultic by nature. For instance we think of the glassy-glassy game and the suggestion that something supernatural is at work when it is played. Before long those who play this game discover that the glass does inexplicable things. So they conclude that some or other invisible factor or spirit is involved.[15] This intrigues them and entices them to seek more *discoveries* in this sphere. This road delivers them into the hands of satanists who pretend that they can offer such people more of these fascinating things. Satanism is then supposed to help its victims to develop unknown skills.

Satanistic placards, books, graffiti, T-shirts, music and numerous occultic objects worn by people are means to propagate this evil cult.

It appears that rock music groups (some of whose members are satanists) contribute substantially in preparing people to accept this cult.[16] Since satanism is focused on lust, the so-called heavy metal music aptly suits this purpose. This music is very loud and rhythmic (carried by loud drum beats) and is accompanied by flashing lights, violent physical movements and screaming. The lyrics are blasphemous whilst Satan is glorified. In addition terrible suggestions are made, such as the practice of patricide.[17]This kind of music serves as a suitable catharsis for the pent-up personal problems, frustrations, surging emotions and desires of some. As the music continues and becomes more intense, a mass hysteria is aroused among large audiences. This causes people to fall into a state of frenzy where they become easy prey to evil influences.

Children and young people, in particular, are the object of the satanistic onslaught. They are obviously still uninformed, curious, receptive to all kinds of influences and still incapable of identifying potentially dangerous in-

14 Op cit, pp 33–35.

15 John and Helen Gardiner declare: "There are many things in the occult that young people especially find fascinating. And the biggest mistake we can make, is to pretend that the occult is just hocus-pocus that doesn't work. It does work, and that's the problem." Op cit, p 80.

They also narrate this testimony of a young lady: "I saw the glass move on its own and spell out messages across an alphabet spread out on the table. I was stunned to realize that there were actually real spirits in the air; I saw the evidence right in front of my own eyes. Op cit, p 21.

16 Cf in this connection an article of Rodney Seale, Satan is openly venerated by this music. *Boodskap van die Basuin*. Sept–Oct 1987, pp 6–9.

17 Cf John and Helen Gardiner. Op cit, pp 117–123.

fluences. Members are usually recruited in satanistic cells amongst school-going children. Those who become interested and look for more *adventure* are then introduced to older satanists.

Those who usually get caught in the snares of satanism are children who grew up in unhappy homes; lacked proper parental guidance; did not receive the necessary love and protection; and the children of satanists. These people regard the interest shown by older satanists in them as the recognition they always craved. People who know nothing of Christ and true Christian fellowship and have no moral anchor are easily carried away by this evil force. This evil degeneration is further facilitated by the promise of free drugs, cheap sex and the special powers with which one would be endowed.

Once under the power of satanism, they are stripped of all moral values which they may have possessed. By exposure to and participation in the most extremely gruesome deeds, they also have to deny God, Christ, the Bible, the church or any religious conviction which they may still have cherished. In addition they also link themselves to Satan by validating a document with their own blood.

Once people have become satanists, they are warned never to break this blood covenant. If they do this, a cruel curse would be pronounced over them. Consequently this is why people who have become disillusioned with satanism are too afraid to turn their backs on it.

1.3.3.3 Satanism as an anti-Christian cult

The above description of satanism is but a contemporary version of what we read in 2 Thessalonians 2:9–12. In this passage of Scripture the Anti-Christ and his influence over man is portrayed.

Satanism is the embodiment of Satan in those who already act as anti-christs. One can hardly imagine how terrible it will be when the anti-Christian force, in total abandon from that which holds it back, will rage unrestrained over the entire world.

Satanism already shows us how the Evil One is exalted as god and how Christ and Christians are despised. Satanists are taught to break up marriages, to destroy Christian institutions and act in all kinds of lawless ways. Children are even persuaded to commit murder.

In contrast we have the gospel. This frees man from destructive forces and lifts him up to become the creature created in God's own image. The converse is true of satanism. People are exposed to the most shocking and inhumane treatment. They are degraded to slaves of the Evil One so that they eventually become robots without a will of their own.

Satanism mocks the sacrifice of Christ on the cross by sacrificing animals and even people. In this way Satan also has his own form of holy communion. During such an occasion adherents of satanism have to eat parts of human bodies and drink the warm blood of a recently slaughtered cat. The most important days on the Christian calendar are used to ridicule, in the most terrible way, the things for which they stand.

1.3.3.4 The deliverance from satanism and its significance for the understanding of demonism

Typical of demonising, the satanist experiences a strange power or presence operating in him by way of an irresistible obsession. It forces him to think, do and say evil things. However, these same people can actually be freed when they turn to Christ in true repentance and confession of sin. We can gratefully say that numerous former satanists are today proclaiming the gospel of Jesus Christ. They are especially zealous to free those still in the power of the Evil One.

Satanism provides us with a clear example of demonising where the thinking and the will of the subject are not necessarily completely eliminated. The same victim who submitted himself voluntarily to Satan in order to serve him can with the help of the Lord and other Christians turn his back on the Evil One and receive deliverance in Christ.

2 DEMONISING AS DEMONOMANIA

As has already been mentioned, the current term for *demonomania is demon oppression, demon obsession or demon possession.* We are dealing here with a psychiatric and, sometimes together with this, a physical condition caused by a demon or demons.

The gravity of this condition can assume varying degrees. At worst a victim's personality is so overpowered, supplanted and disturbed by demonic forces that a strange person and power is revealed in him. An example of this is the demoniac in the country of the Gadarenes (Mark 5:1–15). In order to explain such a serious case in a simple way, we can compare the victim to someone who has been hypnotised by someone else. The hypnotised person would then speak and act in the way instructed by the hypnotist.[18]

2.1 Is there a difference between demonic oppression, obsession or possession and demonomania?

There are theologians who answer in the affirmative. It is alleged that we, on the one hand, have three different biblical expressions.

Firstly there is demonic oppression or being *vexed with a devil* (Matt 15:22), *troubled by unclean spirits* (Luke 6:18; Acts 5:16) and *being under the power of the devil*: oppressed by the devil (Acts 10:38).

Secondly, there are those who live within the realm of the devil, such as *to be with an unclean spirit* (Mark 1:23; 5:2) *to have a devil* (Matt 11:18), and *to have an unclean spirit* (Mark 7:25; Luke 4:33).

Thirdly we have the case where the devil or evil spirits are inside a person's body without these possessing him as, for instance (Mark 9:17, 25), *to have a dumb and deaf spirit*, and (Luke 13:11) *to have a spirit of illness.*

Besides these we have the case of *demon possession.* This is when one is controlled internally by a demon or demons. That is expressed by the word demon-possessed as we find it in the following references: Matthew 4:24; 8:16; 8:33; 9:32; 12:22; 15:22; Mark 1:32; 5:15; 16:18; Luke 8:36, and others.

18 W C van Dam remarks: "In demon possession the same mechanism as that of hypnosis is functioning. During the hypnotic state as well as in the crisis situation of a possessed person, super achievements appear. The end of the trance situation is like coming out from a hypnotic condition. In both instances there is no memory of what had happened. The most simple way to explain the crisis situation of possession, is that the demon has hypnotised somebody and thus controls his conduct." (translation – F.P.M.) Op cit, p 75.

In response to this, we point out the following. Whether we speak about obsession, oppression or possession, one thing is clear, something in the inner person is wrong. If this abnormality is *inside* a person, then the forces which are causing the trouble must also be *inside* the person.

There are no biblical indications that in the one case (obsession and oppression) the demonic force is acting from the outside and in the other case (possession) it is acting from the inside.

Though the Bible uses different words in expressing satanic attacks, deliverance from all of these occur in the same way. In each case deliverance comes by means of the exorcising of the demon. We read of the casting out of demons or devils not only in cases where the word *demon possessed* is used, but also where other expressions are used. Examples are: Mark 1:23–26; 5:2, 12, 13; 7:25, 30; 9:17; Luke 4:33–35.

Furthermore, we see that the word *heal* is used for cases in all categories. Thus we read about *healing* where the term for *demon possession* does not occur, for example Luke 6:18; 13:11–13; Acts 5:16; 10:38. However, we also read of *healing* in cases where demon possession is mentioned, for example: Matthew 12:22; 15:22; Luke 8:36.

Subsequently we should note that demonising is not something spatial, something localised at some place, but of an evil power in relationship with man. It is about a power, a hold upon or an influence which the devil has over a person. Consequently it is to get the evil power out of a person's life that we should understand the notion of casting out or exorcism.[19]

Likewise we should also point out that the idea of *possession* (together with *obsession* or *suppression*) is not reflected in the term used in the Greek text. To be demon-possessed is a concept popularly used to translate the Greek word *daimonizomai*. We, however, want to avoid giving a meaning to *daimonizomai* through translation which does not belong to the word itself. In this way it has also been taken up in the parlance of the Bible. For this reason we choose to talk about *demonising* and *demonomania. Demonomania* signifies insanity due to demonic influence.

2.2 Symptoms which indicate a possible demon mania

According to the biblical record certain symptoms can be connected to demonomania.

2.2.1 Physical symptoms

In this regard the following are mentioned: supernatural power (Mark 45:2–4), blindness, deafness and dumbness (Mark 9:17, 25; Matt 9:32, 33; 12:22), foaming at the mouth (Mark 9:18; Luke 9:39), convulsions

19 Regarding this, Stephen D Swichart remarks: "A careful examination of these passages will disclose to you the fact that no distinction is made between internal activities (supposedly, possession) and external activities (supposedly, oppression). In each instance the demonic work occurs within the person under attack." Op cit, pp 169–172.

In this connection John Wimber and Kevin Springer write: "The demonised suffer under varying degrees of bondage. A demonised person may be tormented by troubling thoughts or may be controlled by a demon (what is commonly inferred from the English term possession'); yet the New Testament would only say that both are demonised'. So referring to all demonised people as being possessed is not Biblical; it is inaccurate and tends only to confuse and emotionally upset people." *Power Healing*, 1990, p 123.

(Mark 1:26; 9:18), gnashing of the teeth (Mark 9:18), falling and rolling on the ground (Mark 9:20; Luke 9:42) and total exhaustion after the demons have departed from such a person (Mark 9:26).

2.2.2 Psychological symptoms

Under this heading we can mention the following: self-infliction of injury (Mark 5:5); suicidal tendencies (Mark 9:22); extremely aggressive behaviour toward others (Acts 19:16); screaming (Mark 1:23; 5:7; Luke 9:39), tearing off clothes (Luke 8:27); supernatural knowledge (Acts 19:15), and a split personality (Mark 1:34; 3:11; 5:9). All these instances suggest that the behaviour of the victim is of a compulsive nature. He acts like one compelled by an unseen power. Such a person can also act completely asocially and hide in desolate places (Luke 8:27).[20]

2.2.3 Religious symptoms

Because we are dealing with evil spirits, we can expect them to reveal themselves in the relationship with Christ and Christian matters. The Bible mentions the following: A voice from a victim who correctly confesses who Christ is and what his apostles are doing (Mark 3:11; Luke 4:34, 41; Acts 16:17; 19:15); the acknowledgement of Christ's presence and the fear of him (Mark 5:7, 10; Matt 8:29; Luke 8:28); a description of the spirits as unclean (Acts 5:16; Luke 4:33; Matt 10:1; Mark 1:23; 3:30); the revelation of resistance and antipathy to everything connected to the Lord and his gospel. The latter is expressed by the words: "What have we to do with you?" (Luke 4:34; 8:28; Mark 1:24; 5:7; 8:29).

In this connection W C van Dam states:

> "Further we want to stress emphatically that if one of the above-mentioned symptoms occurs, it in no way means that we are necessarily dealing with demonising. After all, many of these symptoms also occur in psychiatric profiles. The more symptoms there are present from the mentioned areas, the clearer it will be whether one or more spirits have invaded such a person."[21]

2.3 Demonomania and mental illness

There is a strong similarity between demonomania as described in the Bible and mental illness as known in medical and psychiatric circles. This similarity is especially evident in the cases of the so-called lunatic son (Mark 9:17–19) and epilepsy as known in medical circles. However, similarity still does not in any way imply that all cases of mental illness can be ascribed to demonising.

We first make a general observation in this regard. There are cases where any condition of sickness can be the direct result of sin and even of demonic operations. Most cases of illness, whether physical or mental, are,

20 Those who deal with demonomaniacs mention symptoms like the following: an obsession to harm others or to murder them, unnatural restlessness and anxiety, speaking in a voice which is strange to the speaker, various diseases which cannot be cured by medical, psychological or psychiatric treatment, emotions which do not fit the circumstances, sudden change in personality and character. Cf Jessie Penn-Lewis, *War on the Saints*, pp 140, 141; H A Maxwell-Whyte, *Dominion over demons*, pp 15, 16.

21 Op cit, pp 17, 18.

however, only indirectly connected with sin and the Evil One. People become ill on account of the brokenness and sinfulness of mankind as a whole, as well as on account of the natural circumstances in which they find themselves. Before one can speak of demonomania, there must first be clear signs which distinguish it from ordinary mental illness.

According to Holy Scripture the following could comprise such signs: (a) when the presence of the Lord or his disciples created a situation of crisis for such a person; (b) when somebody in such a condition of crisis reveals supernatural abilities; (c) when somebody in a situation of crisis suddenly behaves in a blasphemous and immoral way; (d) where the demon or demons identify themselves; and (e) where one can discern demonic activities by means of the special gift for the discernment of spirits.[22]

2.4 Demonomania and psychiatric treatment

In this connection the following question presents itself: Can demonomania be healed by scientific means? Or, in other words: Can demons be exorcised by means of psychiatric treatment?

Before answering this question we have to take note of the way medical science effects physical healing. Medical healing includes the utilisation of externally orientated procedures like: (a) harmful organisms and toxic substances in the body are neutralised or eliminated; (b) parts of the body which pose a threat to health are removed; (c) restructuring of the body, such as for example the transplantation of organs; (d) parts of the body are kept in a certain position, for example when a splint is used on a broken leg; (e) deficiencies in the nutrition of the body are supplemented; (f) certain organs and functions of the body are stimulated and so on. In brief it implies that medical science and technology do nothing more than help the body to heal itself.

We are of the opinion that the same principle is valid when mental illness and demonomania are identified. All scientific substances can influence the human personality in such a way that the person can obtain control over himself again, and can once again adapt to society. Help of this nature can also enable a patient (or victim) to neutralise demonic onslaughts. This happens even when somebody does not believe in demons. If this happens, it does not imply that such a person has been freed from evil; but it can at least contribute to a situation where such a person no longer has to be classified as suffering from demonomania.

In James 4:7 we read: "Resist the devil and he will flee from you." In this case the believer can rely on the power of the Spirit of God which is in him. In addition any other help may be provided by the Christian pastorate and counselling. Although the person receives this help, the resistance which he poses against evil is still something that has to come from himself – that is his will and ability to pose resistance. In this connection we may accept that

22 John Wimber and Kevin Springer narrate such a case in connection with a woman: "She would have spells in which she became mute, especially when religious matters were discussed. More dramatically when the name of Jesus or Christ was mentioned, she would immediately go into a trance. On one occasion while in a coma, in desperation, a demon was exorcised and her spells ceased. She subsequently accepted Christ as Saviour and has been well since." Op cit, p 127.

the unbeliever who does not have the support of the Spirit of God can also offer a measure of resistance against the devil. Here we deal with the merciful providence of the Lord. Just as there are certain functions in the body which combat disease and promote physical healing, in a similar sense there are certain factors in the human personality which can do the same with regard to spiritual health. The measure of success achieved by psychiatric and psycho-therapeutic means is determined by the measure in which they stimulate and strengthen these healing and correcting factors. On this basis different healing cults also claim a measure of success.

In answer to the question raised at the beginning, we can say that no means of scientific healing can expel demons. These measures may indeed help victims of demonic bondage to free themselves from this evil hold. In this connection the person can also receive help through Christian counselling provided he finds himself in a reasonably normal condition. In such instances he can be assisted in realising that he is plagued by demons and has to be rid of them. He can also be assisted in the identification of one or other sin or event which may possibly have led to being demonised. Furthermore, he can be encouraged to confess these sins, resist the Evil One, forgive those who harm him and yield himself to Christ as his Saviour.

In connection with demonomania (and even mental illness in general) there are, however, cases which are of such a serious nature that deliverance is only possible by exorcism of demons or through the spiritual gift of healing.[23]

Whether a person has been helped by scientific remedies or by exorcism of demons, true deliverance, however, only takes place when a person dedicates himself to the Lord, and when his *entire* life is controlled by his Spirit. It is then that we see the promise of John 8:36 come into operation: "Therefore if the Son makes you free, you shall be free indeed." The person who lives in this freedom need never be afraid that the Evil One will get a hold on his personality and thus become a demon maniac.

2.5 The difference between demonising in general and demonomania

We have already pointed out that demon mania is a form of demonising. The question now is how this differs from other forms of demonising.

A watertight distinction is just not possible; there is simply too much overlap for this. On the other hand, there have to be certain criteria according to which a distinction between demonomania and other forms of demonising must be made. We refer to some of these.

The other cases of demonising relate mainly to the personal, direct sinful relationship between people and the Evil One. It is a case of people purposefully identifying themselves with that which effects demonising. This

23 W C van Dam mentions instances which confirm these views. Op cit, pp 62–64. Kilian McDonnell remarks: "Can we not assume that beneath the outward appearances of illness and psychoses, Jesus sometimes perceived a malignant force at work whose purpose was ever to bring sickness where there was health, division where there was life?" *Presence, Power, Praise,* Vol 1, p 244.

identification with evil may indeed lead to demonomania, yet this is not inevitable.

In demonomania the emphasis is more strongly on the sick personality resulting from demonic operations. For this reason exorcism is mentioned together with other kinds of healing. Deliverance from demonomania is consequently described in the Bible also as a healing (Matt 12:22; 15:22–28; Luke 6:18,36; Acts 5:16; 10:38).

As a rule, other demonised people reveal a reasonably integrated personality. In addition their character is more stable. The evil which they reveal is something which is always typical of them. Their entire lives are so integrated around evil that they do not necessarily show serious inner conflicts and personality disorders. Many of those who deal with such people will not readily discern that there is something peculiar about them.

The demonomaniac, by contrast, reveals the typical profile of illness where one has the oscillation of a reasonably normal and a crisis condition. One moment such a person can appear normal; and the next moment he may again act insanely. This is also reflected in the moral nature of his behaviour. One moment he will behave in a civil, well-mannered, morally sound and even pious manner. However, when he enters a crisis situation, he shows his immoral behaviour, for instance by swearing and blaspheming.

In other cases of demonising the particular person deliberately and consciously does evil things. He remembers what he did and assumes responsibility for it.

In the case of demonomania the victim is usually unaware of what he did in the crisis situation. When others tell him about everything he did and said, he himself cannot believe it. Often his behaviour in that condition was the opposite of the kind of behaviour he would have liked to reveal. The fact that such a person's behaviour betrays an inner conflict may actually mean that he is trying to resist a strange power.

The deliverance of other demonised persons occurs when they hear the gospel. When a person accepts the gospel, he shows sorrow, repents and breaks with his old life while receiving Christ as his Saviour. However, where there are also demonic elements present, exorcism has in fact to take place.

Other cases of demonising, such as spiritism, occultism, satanism, and so on, pose a serious threat to the church of Christ. As we have already seen, this is an enticing power which endeavours to bring others under its influence.

Demonomania is different. Because this is a state of illness, it poses none of the above-mentioned threats.The insane behaviour revealed by such a person in his crisis situation will rather repel than attract people.

2.6 Can a believer be a demonomaniac?

When the believer reads about the cases of demon mania in the Bible he may indeed ask himself: Can something like this also happen to me?

Before we try to answer this question, let us first consider the fact that a believer can indeed be mentally ill. There may be many reasons for this, such as: neurological and physiological disturbances, hereditary factors, traumatic experiences, and so on. The fact that such a person is a believer can indeed contribute to overcoming these personality problems. However, a person will not always remain spiritually and physically healthy because he is a believer. The question whether a believer can have spiritual and other disorders as a result of demonic bondage in his life, is, however, not so easy to answer from a dogmatic point of view. This question has been taken to extremes in theological circles, precisely because theologians embrace conflicting views.

2.6.1 Those who say that demons can be exorcised from believers

To begin with, there are various prominent theologians and missionaries who claim that they can quote many instances where believers were delivered by means of demon exorcism from the one or other bondage or behaviour disorder.[24] Although we have no right to doubt the testimony of many respected theologians and missionaries, we are not allowed to build doctrines on what has happened in practice. First and foremost, dogmatic statements have to emanate from Scripture. Sometimes Scripture makes no direct pronouncements about a certain matter. In such a case any proposition should not clash with biblical doctrine, but should rather link up with it.

At the same time we can in fact expect that our interpretation of biblical truths should also be realised in everyday life. That which God presents to us in Holy Scripture should after all be the same as that which we encounter in practice. If there is a conflict between theory and practice then something must be wrong. It may imply that our doctrinal interpretation of what we perceive in practice is incorrect. It may also imply that our interpretation of what we read in the Bible is incorrect. By referring to this last possibility, we by no means want to suggest that practice, apart from the Bible (or even in place of the Bible) can be the basis for our confession. What we do say, however, is that our interpretation of the biblical truth may be superficial, one-sided and even prejudiced. When there is a clash between our interpretation of the Bible and practice it should encourage us to reconsider our interpretation of Holy Scripture as well as our own observation from practice.

24 Maxwell-Whyte mentions various instances where demons were cast out of believers. *Dominion over demons*, pp 17, 18. C Peter Wagner communicates the following: "In my research I have noticed several things. For one, almost all those who themselves are actively involved in a ministry of exorcism or deliverance, affirm that Christians can be demonized. For another, I have discovered that several Christian leaders have changed their opinion on this matter in recent years. But the changes I have observed have all been in the same direction, namely, from once denying that demons can harm Christians to now affirming that they can and do." *How to have a healing ministry*, p 190.

Prof Fred Dickason communicates the following findings: "I have encountered from 1974 to 1987 at least 400 cases of those who were genuine Christians who were also demonized. I am not gullible or easily convinced ... I would not claim infallible judgement, but I know the marks of a Christian and the marks of a demonized person. I might have been wrong in a case or so, but I cannot conceive that I would be wrong in more than 400 cases. I have referred many cases to others, such as pastors, psychologists, and psychiatrists, who confirmed my diagnosis." *Demon Possession and the Christian*, 1987, p 175. Similar notions are also expressed by W C van Dam, op cit, pp 19, 39, 121, as well as Merrill F Unger, op cit, pp 113, 184.

2.6.2 Those who maintain that it is impossible that demons can be exorcised from believers

On the other hand certain theologians declare that it is quite impossible and unscriptural for a true believer to still have demons within himself. In order to substantiate this view reference is made to several matters. We base our discussion on views as they appear in the theological volume *Power Encounter*.[25]

To begin with, reference is made to the fact that lexicons translate the Greek concept daimonizomai with to be possessed by a demon or demons. To this the idea is added that one cannot be possessed by God and a demon at the same time.

Concerning this viewpoint, we can only say that no dogmatic conclusions can be drawn from the meaning ascribed to the translation of a Greek word. As has already been mentioned, the idea of possession is linguistically not part of the original word.

It is also pointed out that one's body is the temple of the Holy Spirit. In addition Scripture teaches that Christ overcame the previous owner of our lives, namely the devil, and made his abode with us. God and the devil can surely not dwell in the same house!

Biblical images such as temple and house may, however, not be extended any further than the main idea they wish to convey. This deals mainly with one's responsibility towards the Holy Spirit. Furthermore it deals with Christ's victory over the Evil One. If we want to interpret these images any further, we end up with another problem. If it deals with the fact that Christ does not dwell in the same house as the devil, then it must also imply that He will not dwell in a house where demonic dispositions are still revealed. Paul points out that the following are still found in the lives of some believers: hatred, jealousy, fear, uncleanness, adultery and similar matters (Eph 4:31; Jas 1:19–21; 1 Pet 2:1, 16). To counter that these things are the works of the flesh and not the revelation of demonic operations is biblically incorrect. All anti-God characteristics revealed through the flesh come from the devil (1 John 3:8). To further say that the sinful nature or our carnality is under the blood of Christ and therefore not accessible to demonic influence is simply not biblically substantiated. Nowhere does Holy Scripture teach that the purification or annihilation of our sinful nature takes place through the blood of Christ. What it does teach is the mortification thereof. We are admonished to consider this nature as crucified and dead (Rom 6:1–11; 7:15–18) and only as far as that happens, it affords no gateway for evil to enter.

Furthermore, it is mentioned that there are missionaries who declare that they have never driven any demons from believers. On the basis of this it is then said that it is impossible for demons to inhabit the lives of believers.

The error in this kind of argument amounts to the so-called *argumentum e silentio*: because you do not observe something, therefore it simply does

25 Elmer Kirsch, Power in Biblical Christianity, pp 60–68, Can born-again believers be demon possessed? pp 269–277. *Power Encounter, A Pentecostal Perspective*. C B C Press Publishers, Springfield.

not exist! The fact remains, however, that although some missionaries or pastors have not encountered this, others have.

It is further claimed that believers were never the victims in any of the instances where the Bible speaks of demon exorcism. In addition it is said that the gift of the discernment of spirits (1 Cor 12:10) was exclusively given with a view to unbelievers.

The fact is, however, that nowhere does the Bible say anything directly or indirectly about the fact that demons can only be driven from unbelievers.[26] The Bible also does not say that the gift of the discernment of spirits was only given with a view to unbelievers.

Strong emphasis is laid on Scripture passages which state that in Christ we have complete victory over the Evil One. The following references are mentioned: John 8:36; Romans 6:20; 8:15; 1 Corinthians 15:57; Galatians 3:13; Colossians 2:15 and 1 John 4:4.

These texts do indeed refer to the total victory which one can and should have in Christ. However, they do not imply that all Christians experience that complete victory simply because they have become believers. The fact remains that the Bible also talks about the believers' struggle with the Evil One, and against the evil forces (Eph 6:11, 12; Heb 12:4). The texts cited, thus speak of the danger which the devil still poses to the believers (Luke 22:31; 1 Cor 10:21, 22; 2 Cor 2:10; 1 Tim 3:6, 7; 2 Tim 2: 25, 26; 1 Pet 5:8).

Then there are those who maintain that Christians can be demonised, but that such Christians are not really born again and have also not learnt to know Christ as Lord of their lives.[27]

This is, however, merely a statement with no substantial proof. Before saying this, one has to determine precisely the spiritual condition of all these people.

2.6.3 Demonomania as a demonic bondage

To arrive at a possible resolution of this difference of opinion, we have to first determine the meaning of daimonizomai. If it means to be possessed by a demon or demons whereby a person's life belongs to the Evil One and is under his direct control, then it is evident that a true believer cannot be a demonomaniac.

It is, however, a totally different matter when we consider demonomania as referring to mental illness or a personality disorder caused by demonic bondage. We speak here of bondage because as we have seen, it is about a state of illness. In the case of the unbeliever who has not yet experienced the quickening operation of God's Spirit, there can in fact be such an evil grip on his life whereby everything he does is demonic. However, this need

26 After having pointed out that the Bible does not distinguish among *oppression, obsession* and *possession*, Swichart remarks further: "And finally, there is no attempt to distinguish between Christian and non-Christian demonization. Everyone is a possible subject for these assaults." Op cit, p 172.

27 This is the viewpoint of Edward N Gross, Miracles, Demons and Spiritual Warfare, pp 164, 165.

not be the case; the demonic bondage may only affect a part of his life.[28] The believer whose spirit is reborn, and who is united and filled with God's Spirit can obviously not be under the total control of the Evil One. He may have bondages in his personality, but as regards his spirit, he is complete and in God's hand.

This view proceeds from the teaching about the two natures in the life of the believer. In Romans 7:19–24 Paul refers expressly to the two opposing laws and forces at work in him as a believer. In his spirit he enjoys a genuine relationship with God. In that sense he is complete and is seated with Christ in the heavenlies (Eph 2:6). However, with regard to his carnal nature he is still subject to sin. Through this nature the devil gains a foothold in the life of the believer. By means of this nature, demonic bondage may realise in the life of the believer. According to testimonies, such bondages may even have existed before such a person came to faith and salvation. These are usually linked to a person's previous involvement in occultic acts.

When demons are driven from such a person, it implies that the hold of the Evil One on his life is destroyed. Whether this was indeed a case of demonising can only be confirmed when such a person has been truly freed and can testify to it. What we try to say is that though the spirit of a true believer is controlled by God alone, he can still suffer from psychological bindings caused by demons. Such binding can result in some kind of demonomania.

At this stage we have to state very clearly that any form of demonomania in the life of the believer is an exception, and is absolutely contrary to the life which such a person has in Christ. Such a person cannot and should not rest until he is freed from any binding of the Evil One.

There is no guarantee that demons cannot enter a person just because he has once come to faith. Our protection against evil onslaughts does not reside in a one-time experience, but in a relationship with God. For this reason believers should always walk in the Spirit (Gal 5:16), that is, they should remain in fellowship with God. Only then will the sinful flesh be prevented from gaining an upper hand over a believer and creating an opportunity for Satanic attacks. He who continues in a loving relationship with God does not have to fear any evil force; he will only experience that in Christ he is more than a conqueror.

Before proceeding to the exorcism of demons, we must first make very sure whether it is a case of demonising. One cannot indiscriminately proceed with exorcism, simply because there is evidence of one or other sin, a bad habit, some irritating or perverse behaviour. He who wants to exorcise demons where there are no demons, not only renders the entire mat-

28 Kilian McDonnell also does not distinguish between demon possession and demon oppression, and declares: "To be oppressed by a demon may be a Christian's experience in some area of his life, from which he has barred Jesus Christ as Lord, or at such a point at which he has consciously opened up his life to the influence and sway of demonic powers. Where the rule of Christ is held at arm's length, a Christian may well find himself powerless, and driven helplessly by fierce passions and obsessions. Where a person's life has been given access to forces of evil, these forces may mount a frightening attack on him, even after he has committed himself to Christ, and they may try to drive him to destruction." Op cit, p 578.

ter ludicrous, but can also cause trauma and confusion in the life of those he is supposed to help.

3 CHRIST AND DEMONS

Everything we read in the New Testament about the devil and demons has to be understood in the context of Christ's dominion over all evil forces.

3.1 When evil forces act against Christ and his church

We read in the Bible how the devil and his demons acted against the Son of God in the time of his physical sojourn on earth. It is precisely in connection with Christ and the true believers that demonomania is often mentioned.

Concerning the unbelieving world, the devil and his angels apparently feel no need to reveal themselves personally and directly. They rule already over sinful humanity (1 John 5:19). In addition, we should mention that certain theologians and churches in the Christian tradition deny the existence of the devil and his demons. These quarters obviously pose no threat to the evil forces. It becomes easier for the Evil One to remain confidently in the background. From there he can launch his deceiving and destructive onslaught through people.

However, it is different when the devil and his demons encounter Christ and true believers. It is then that they come to the fore in a more personal and direct way by demonising people.

Holy Scripture makes it clear that there were exceptional demonic operations at the first coming of Christ to earth. We read how evil operations were put into practice in order to murder the Christ-child (Matt 2:1–18). In Matthew 4:1–11 we see how the devil personally tried to tempt Christ to deter him from accomplishing his work of salvation. We also see how Satan, through the Jews (and the Roman authorities) resisted Christ's ministry. Furthermore, we detect how Christ's actions and power were challenged by the numerous demoniacs who encountered Him on the way. Eventually the evil onslaught reached such proportions that nothing could prevent them from crucifying Christ.

The irony of this onslaught is that the devil well knew what lot would befall him and his angels because of Christ's death on the cross. Christ's death on the cross did not imply defeat but victory.

It is all important to note that the victory over Satan and his angels won by the Lord Jesus on the cross was not accomplished for Christ Himself. He was and always is in control of everything (Col 1:15–17). His victory was for the sake of all humanity. It is the redeemed who became more than conquerors through Christ. The devil's strategy was to use God's righteousness against his own creation. He corrupted creation through sin and in this way brought it under God's righteous judgement. His claim that all people should be condemned like himself was nullified. By the death on the cross of his Son, God brought about reconciliation, salvation and eternal life. In this process He also exposed the Evil One, made a spectacle of him as the enemy of his own Creator and Sustainer. Thereby the eternal judgement of the devil and his angels was sealed (Col 2:13–15).

As one who knows that his time is short and the day of his judgement is inevitably drawing near, the devil in the meantime rages with all his angels against the church of Christ (Rev 12:12–17). Apparently he wants to prevent any deeds which would stop people from being cast into eternal damnation along with him. In this regard we discover another irony – an irony which testifies to God's overall dominion. While the Evil One persecutes the church of Christ, he is actually strengthening the Christian testimony and causing more people to be saved (Phil 1:12–14).

Just before the return of Christ, the Evil One will be given free reign for a while. Then he will try to destroy the Christian witness on earth through the Antichrist and the anti-Christian spirit. At that time there will be many more direct demonic revelations through people. That the devil will do: "...with all power, signs, and lying wonders, and with all unrighteous deception among those who perish ..." (2 Thess 2:9–10). When Christ appears all evil forces will finally be subjected to Him (1 Pet 3:22; Eph 1:20–23). The devil and his angels together with all ungodly people will then be cast into perdition or hell. Consequently this is the place that has been prepared for the devil and the angels (Matt 25:41; 2 Pet 2:4, 9; Jude 6). This will mean the end of the devil and his demons.

3.2 Christ and the exorcism of demons

It is worthwhile to note the way in which Christ acted toward demons in his earthly ministry. Firstly, we see that his behaviour towards the demonised was different from his behaviour towards the physically ill and disabled; He was deeply sorry for the latter. Sometimes He went to them and began to converse with them. Often He touched them and healed all those who were physically ill.

When He encountered demonised people, He did not begin to converse with them. He directly confronted the demon or demons. He approached every demon abruptly, and with contempt and commanded it with a rebuke to depart from such a person (Matt 8:16; 17:18). No resistance on the part of the demons is mentioned. The contrary is true. Both in the case of Christ's own ministry, as well as when He sent out his disciples endued with special power, demons had to flee without any ado (Luke 10:17). Demons shuddered (Jas 2:19) and screamed with fear in the presence of Christ (Matt 8:29; Mark 5:6,7). They also begged Him not to send them to the nether world; rather they wished to beset a herd of pigs (Luke 8:30–33; Mark 5:10–13).

The casting out of demons is described in terms of clear observable events in Holy Scripture. It is mentioned that when a demon left a victim, he would scream through such a person, make him convulse and cause him to fall to the ground as if dead (Mark 9:26).[29] However, it can also happen that demons can leave a victim without any visible or audible manifestations. After the casting out of demons the victim was free and as normal as anybody else (Matt 9:32, 33; Luke 8:35). This all signifies Christ's dominion over demons.

29 Contemporary testimonies of exorcism also speak of vomiting by the victim. Some count the number of devils leaving by the instances such a person vomits. Such phenomena must not be understood as if the devils were resident in a victim's belly. It must be rather viewed as the victim's bodily reaction to the departing demons.

Christ also gave this power to drive out demons to his disciples. Besides proclaiming the gospel and healing the sick, they also had to cast out demons. Subsequently this took place from the time of the apostles throughout the history of the church until today. There were indeed times when it seemed that miraculous healings and the exorcism of demons disappeared from the Church. In times of spiritual revival, however, the church experienced these things again.[30] In contemporary times there have been numerous reports from different quarters of demon exorcism.[31] People usually refer to the driving out of demons as exorcism. On account of the occultic origin of this word and its association with occultism, it is inadvisable to use this term in Christian circles.

It is also clear that the name of Jesus Christ should not be used like some kind of magic formula when driving out demons. This is clear from the events recorded in Acts 19:13–16. One can only exercise authority over the devil and his demons when one is under the authority of God and when one acts in the power of his Word and Spirit.

According to New Testament reports, the driving out of demons takes place immediately. In this regard it differs from some present-day testimonies where an entire process takes place before a person can be freed from demons.

When somebody has been freed from demons, he still has to guard against further evil deception. At the same time he must dedicate his entire life to God and his church. The devil will always try to regain lost territory. In this battle a person needs the loving and caring fellowship of other believers.

30 Cf Kenneth Scott Latourette, op cit, p 105; Maxwell-Whyte, op cit, p 26; W C van Dam, op cit, p 14.

31 The word is derived from the Greek exorkisô which means literally: to bind by an oath. According to *The Consolidated-Webster Encyclopedic Dictionary exorcise* is described as: "to expel or cast out by conjurations, prayers and ceremonies." p 265.

Bibliography

Aalders, G. Ch. 1932. *De Goddelijke Openbaring in de eerste drie hoofdstukken van Genesis.* Kampen: J.H. Kok.

Barth, Karl. 1957. *Kirchliche Dogmatiek Vol. 3,1.* Zurich: Evangelischer Verlag U.G. Zollikon.

Barth, Karl. 1958. *Kirchliche Dogmatiek Vol. 3,3.* Zurich: Evangelischer Verlag U.G. Zollikon.

Barth, Karl. 1959. *Kirchliche Dogmatiek Vol. 4,4.* Zurich: Evangelischer Verlag U.G. Zollikon.

Bauernfeind, Otto. 1972. *Theological Dictionary of the New Testament. Vol. 3.* Grand Rapid: Eerdmans Publishing Co.

Bavinck, H. 1928. *Gereformeerde Dogmatiek Vol. 1.* Kampen: J.H. Kok.

Beek, M.A 1951. *Aan Babylons Stromen.* Amsterdam: Uitgewers-Maatschappij "Kosmos".

Berkhof, H. 1952. *Christus en de Machten.* Holland: Nijkerk.

Block, David, L. 1992. *Our universe: Accident or design?* Johannesburg: University of the Witwatersrand.

Brown, Driver and Briggs. *Hebrew and English Lexicon.*

Brunner, E. 1960. *Dogmatik Band 1 and 2.* Zürich, Switzerland: Zwingli Verlag.

Bultmann, R. 1941. *Neues Testament und Mithologie.* Hamburg, Volksdorf: Herbert Reich.

Calvyn, J. *Institutie* (Sizoo). Holland; Delft: W.D. Meinema.

Capps, C and A. 1984. *Angels.* Tulsa: Harrison House.

Catherwood, F. Cf. "What the scientists say about evolution". *Dedication.* Vol.2 No. 2. Slough, England: Gospel Tabernacle.

Coder, M. and Howe, G.F. *The Bible, Science and Creation.* London: Victory Press.

Crick, F. 1969. The origin of the genetic code. *Journal of Molecular Biology.* Vol. 38.

Darwin, C. 1859. *On the origin of species by means of natural selection.* London: John Murray.

Davidson, B. *The Analytical Hebrew and Chaldee Lexicon.* London: Samuel Bagster and Sons.

Davidson, G. 1967. *A Dictionary of Angels.* New York: Free Press.

Denton, M. 1986. *Evolution, a theory in crisis.* Bethesda, Maryland: Adler and Adler Publishers.

Duffield, G.P. and Van Cleave, N.M. *Foundations of Pentacostal Theology.* Los Angeles: L.I.F.E. Bible College.

Du Toit, S. 1968. *Bybel, skepping, evolusie.* Pretoria: Unie Boekhandel.

Duyvené de Wit, J.J. 1963. *The Palaeontological Record and the Origin of Man.* Bloemfontein: University of the Free State.

Dijk, K. *Korte Dogmatiek.* Kampen: J.H. Kok.

Eichrodt, W. 1961. *Theologie des Alten Testament.* Stuttgart: Ehrenfried Klotz Verlag.

Foester, W. *Theological Dictionary of the New Testament. Vol. 1.* Grand Rapids: Eerdmans Publishing Co.

Gaebelein, A.C. 1924. *The Angels of God.* New York: 456 Fourth Avenue.

Gardiner, J.H. 1990. *Satanism: The seduction of South Africa's youth.* Cape Town: Struik House.

Geyser, A.S. 1946. *Die geslagsregister van Jesus Christus volgens Matt. 1:1–17 en Luk. 3:23–38.* Pretoria: University of Pretoria.

Graham. 1976. *God's Secret Agents.* London: Hodder and Stoughton.

Gribben, J. and Rees, M. 1991. *Cosmic Coincidences.* London: Black Swan.

Grosheide, F.W. 1950. *Bijbelse Encyclopedia.* Kampen: J.H. Kok.

Grosheide, F.W. 1954. *Het Heilig Evangelie volgens Mattheus.* Kampen: J.H. Kok.

Gross, E.N. 1990. *Demons and Spiritual Warfare.* Grand Rapids: Baker Book House.

Gschwend, J.R. *Light rays in a dark world.* Pretoria: All Nations Gospel Publishers.

Gunton, C.E. 1991. *The Promise of trinitarian Theology.* Edinburgh: T & T Clark.

Heyns. J.A. 1975. *Nederduits Gereformeerde Teologiese Tydskrif,* Sept. 1975. Stellenbosch.

Hunter, C. and Frances. 1979. *Angels on Assignment, as told by Ronald Buck.* Kingwood: Hunter Books.

Johansen, D. and Edey Maitland, L. 1981. *The Beginning of Humankind.* Abridged for the Reader's Digest.

Johnson, P.E. 1991. *Darwin on Trial.* Illinois: Intervarsity Press, Downes Grove.

Keil, C.F. and Delitzsch, F. *Isaiah, Commentary on the Old Testament.* Grand Rapids: Eerdmans Publishing Co.

Kerkut, G.A. *The Implication of Evolution.* London: Pergamon Press.

Kirsch, E. *Power in Biblical Christianity. Power Encounter, a Pentecostal Perspective.* Springfield: CBC Publishers.

Kittel, G. *Theological Dictionary of the New Testament. Vol. 1.* Grand Rapids: Eerdmans Publishing Co.

Koch, K. 1961. *Between Christ and Satan.* Grand Rapids: Kregel Publications.

Koehler and Baumgarten. *Lexicon in Veteris.* Testament Libras.

Larkin, C. 1918. *Dispensational Truth.* Philadelphia.

Latourette, K.S. 1955. *A History of Christianity.* London: Eyre and Spottiswoode.

Leakey, L.S.B. 1960. Finding the world's earliest man. *National Geographic,* 118 (3).

Leopold, H.C. *Exposition of Genesis, Vol. 1.* Grand Rapids: Baker Book House.

Lindsay, G. *Ministry of Angels.* Dallas: Christ for all Nations Publishing.

Luther, Martin. Werke 56, 331, 27.

Maxwell-Whyte, N.D. *Dominion over Demons.* Toronto, Canada.

Matthew-Henry. *Commentary on the whole Bible. Vol. 1.* London: Fleming H. Revell Co.

McDonnell, K. 1980. *Presence, Power, Praise. Vol. 1.* Collegeville, Minnesota: The Liturgical Press.

Möller, F.P. *Words of Light and Life. Vol. 1.* Pretoria: J.L. van Schaik Publishers.

Moore, R. 1964. *Evolution. Life Nature Library.* Netherlands: Time-Life International.

More, L.T. *The Dogma of Evolution.* Quoted in Dedication, Vol. 9. Slough, England: Gospel Tabernacle.

Morris, L. 1971. *Commentary on the Gospel of John.* Grand Rapids: Eerdmans Publishing Co.

Nevius, J.L. *Demon Experiences in many Lands.* New York: Fleming H. Revello.

Nooidtzij, A. *Korte verklaring der Heilige Schrift.* Kampen: J.H. Kok.

Payne, J. B. 1962. *The Theology of the Older Testament.* Grand Rapids: Zondervan.

Penn-Lewis, J. 1983. *War on the Saints.* Fort Washington: Christian Literature Crusade.

Richardson, A. 1969. *Angel, Angelology – A Dictionary of Christian Theology.* London: SCM Press.

Ridderbos, H. 1959. *Aan de Romeinen. Commentaar op het Nieuwe Testament.* Kampen: J.H. Kok.

Rush, W.H. 1972. *Darwin, Evolution and Creation.* St Louis: Concordia Publishing House.

Sauer, E. 1962. *The King of the Earth.* Grand Rapids: Eerdmans Publishing Co.

Seale, R. 1987. Article in *Boodskap van Basuin,* Sept/Oct.

Sunderland, L.D. 1988. *Darwin's Enigma, Revised Edition.* Santee: Master Book Publishers.

Surburg, R.F. *Darwin, Evolution and Creation.* St Louis: Concordia Publishing House.

Swihart, S.D. 1979. *Angels in Heaven and Earth.* Plainfield: Logos International.

Thayer, J.H. 1958. *A Greek-English Lexicon of the New Testament,* being Grimm's Wilke's *Clavis Novi Testamenti.* Edinburgh: T & T Clark.

The Consolidated Webster Encyclopedic Dictionary. 1947. Chicago: Consolidated Book Publishers.

The Meaning of the glorious Koran. 1958. New York: The New American Library of World Literature.

The New English Bible. 1970. England: Oxford University Press.

Tillich, P. 1968. *Systematic Theology II.* Digswell Place.

Tolbert, K.E. 1992. Satanism in America. *Charisma,* Oct.

Unger, M.F. 1974. *Demons in the World today.* Wheaton, Ill: Tyndale House Publishers.

Van Dam, W.C. 1974. *Demonen eruit in Jezus' Naam!* Kampen: J.H. Kok.

Van Niftrik, G.C. 1953. *Kleine Dogmatiek.* Nijkerk: Callenbach Uitgever.

Von Rad, G. *Theological Dictionary of the Old Testament, Vol. 1.* Grand Rapids: Eerdmans Publishing Co.

Westermann, C. 1974. *Biblischer Kommentar altes Testament. Genesis Kapitel 1–11.* Neukirchin-Vluyn: Neukirchener Verlag.

Wilder-Smith, A.E. 1974. *Man's Origin, Man's Destiny.* London: Marshall, Morgan and Scott.

Williams, J.R. 1988. *Renewal Theology, Vol. 1.* Grand Rapids: Academic Books.

Wimber, J. and Springer, K. 1990. *Power Healing*. London: Hodder and Stoughton.

Zimmerman, P.A. 1946. Darwin, Evolution and Creation. *Reader's Digest*, December 1946.